D0334820

Applied Probability

By W. A. Thompson, Jr.,
University of Missouri

Designed for the first course in probability taken by advanced undergraduate and graduate students of mathematics, engineering, and statistics, this text focuses on probability in its applied context. While theory is developed from the beginning, emphasis is placed on the relation of theory to application. Numerous, diversified examples clarify the material which includes information theory, survival functions, the kinetic theory of an ideal gas, and Fechnerian Psychophysics.

In addition, there is a treatment of the multivariate exponential distribution and of the multivariate normal. The chapter "Limiting Extreme Value Distributions" makes an expository contribution to the theory and contains many details not available in other textbooks. Stochastic processes are discussed in a preliminary way and the book is a natural prerequisite for a course in the applied aspects of that subject.

ABOUT THE AUTHOR

W. A. Thompson, Jr. received a B.S. in Mathematics and Statistics at the University of Illinois and his Ph.D. in Mathematical Statistics at the University of North Carolina. He has taught at the Virginia Polytechnic Institute, the University of Delaware, Florida State University, and is presently on the faculty of the University of Missouri.

Dr. Thompson has headed the Computing Branch of the U.S. Air Defense Board, acted as consultant to the Office of Ordinance Research, and as consultant to the Dupont Corporation. Currently, he is the recipient of a National Science Foundation Grant for research in Statistical Theory.

Applied Probability

INTERNATIONAL SERIES IN DECISION PROCESSES

INGRAM OLKIN, Consulting Editor

A Basic Course in Statistics, 2d ed., T. R. Anderson and M. Zelditch, Jr.
Introduction to Statistics, R. A. Hultquist
Applied Probability, W. A. Thompson, Jr.
Elementary Statistical Methods, 3d ed., H. M. Walker and J. Lev
Reliability Handbook, B. A. Koslov and I. A. Ushakov (edited by J. T. Rosenblatt and L. H. Koopmans)
Fundamental Research Statistics for the Behavioral Sciences, J. T. Roscoe

FORTHCOMING TITLES

Introductory Probability, C. Derman, L. Gleser, and I. Olkin
Probability Theory, Y. S. Chow and H. Teicher
Statistics for Business and Economics, W. L. Hays and R. L. Winkler
Statistical Inference, 2d ed., H. M. Walker and J. Lev
Statistics for Psychologists, 2d ed., W. L. Hays
Decision Theory for Business, D. Feldman and E. Seiden
Analysis and Design of Experiments, M. Zelen
Time Series Analysis, D. Brillinger
Statistics Handbook, C. Derman, L. Gleser, G. H. Golub, G. J. Lieberman, I. Olkin, A. Madansky, and M. Sobel

Applied Probability

W. A. Thompson, Jr.
University of Missouri

NORTHWEST MISSOURI
STATE COLLEGE LIBRARY
MARYVILLE, MISSOURI

HOLT, RINEHART AND WINSTON, INC.
New York · Chicago · San Francisco · Atlanta · Dallas
Montreal · Toronto · London · Sydney

123084

Copyright © 1969 by Holt, Rinehart and Winston, Inc.
All Rights Reserved
Library of Congress Catalog Card Number: 79–77815
AMS 1968 Subject Classifications 6001, 6090

SBN: 03-079555-9

Printed in the United States of America

1 2 3 4 5 6 7 8 9

519.1
T48a

TO MY WIFE

Who Understands without Understanding

PREFACE

This book is intended as a "liberal arts" text in probability. It is not a "theory of probability," the examples are too extensive for that purpose. But neither is this a "probability in . . . " where the blank is to be filled with a single area of application; the examples have purposely been chosen from a wide range of applied fields.

We find that some students initially profit by a treatment emphasizing breadth rather than detail, though of course both approaches are important. But the theory and several areas of application are already covered, at various levels, by a number of excellent texts, although the instructor wishing to emphasize perspective is severely limited. Thus the theme of our book is *probability in its applied context.*

Our point of view is well summarized in *The Mathematical Sciences: A Report,* National Academy of Sciences, 1968:

> There should be mathematics courses, available to all, that stress the heuristic process by which mathematical models of scientific questions are arrived at and that emphasize strongly the character of mathematical thought that leads to results from which one can infer the answers to such scientific questions. This is not the same detailed material that most efficiently leads young mathematics students to the frontiers of professional mathematics, and it requires attitudes not necessarily compatible with the attitudes of such mathematics courses.

However, nothing settles a scientific controversy quite as well as the precise statement and logical derivation of conditions under which a certain result will hold; that is, a theorem and its proof. Applied probability is very much like probability except that the emphasis is changed. The subject matter that is central to the theory is also central to the applications; however, the implications of the theorems become more important than their proofs. We are not attempting to write a text for a new academic subject, merely to provide an alternative way of teaching a course that is already offered at most universities. The course that we have in mind is beginning probability at the senior-beginning graduate level.

This book has been developed from course notes prepared at the University of Delaware and at the Florida State University over a period of seven years. The students for these courses have been in mathematics, engineering, and statistics; they have been seniors and beginning graduate students with a minimum mathematics background in advanced calculus. Some of our material is advanced calculus which tends to be omitted from a modern course in that subject. For instance, Chapter 4 contains an appendix on the gamma function. Much of our discussion is elementary and can be used as a review for students with an advanced calculus background. This is particularly true of Chapter 2 on Discrete Probability Measures. But students seem to appreciate the review and the elementary material is needed to make the probability treatment complete. In our experience students with the advanced calculus background have found this material to be deceptively challenging. They are not used to the probabilistic way of thinking and they have never had to marshall their entire mathematics background at one time to the extent needed in studying probability.

The instructor of a one-term course will need to do some picking and choosing of the material that he will cover. Material that can be omitted at a first reading, without loss of continuity, is marked with an asterisk. Chapters 1, 2, 3, 4, and 7 are basic to any course; Chapters 9 and 11 will then round out the term. This will provide a first course in probability, with an introduction to stochastic processes, for the student interested in applications. From the beginning we discuss elementary stochastic processes without using the name; then in Chapter 11 we focus on the process point of view by discussing two examples that have proved useful in applications. We believe that the systematic treatment of stochastic processes is more appropriately included in a second course. There are now many texts on stochastic processes that require as a prerequisite some knowledge of basic probability. We are attempting to meet that requirement for the student of applied probability.

A special feature of the book is that an effort has been made to inconspicuously place the various topics in their historical context. Famous problems have been included by their historical names and short sketches of important probabilists have been included.

All of the material in this book has been successfully presented in class but some of it is not exclusively for that purpose. Chapters 5 and 6 present detailed applications of probability to the physical and social sciences, respectively.

Chapter 5, The Kinetic Theory of an Ideal Gas, presents Maxwell's original work with only slight changes in order to conform to the notation of the rest of the book. This work appeared over one hundred years ago but it is scientifically important today. We find it interesting to see how a really first rate scientist outside our field looks at probability.

Chapter 6 presents classical material which is important in psychology. We hope that we have made some original contribution toward a modern view of these classical results.

Chapters 8 and 10 are theoretical in nature. Chapter 8 attempts to make explicit the relation between probability as it is carried out in abstract theoretical work, and probability as it is conducted in applications, using the tools of ordinary algebra and calculus. This relation is not completely obvious as the proofs of Theorems 8.1 and 8.2 will show.

Chapter 10, on limiting extreme value distributions, is more rigorous than any other chapter in the book. We are attempting to fill in details that we were unable to find in the literature.

A few guidelines on notation will be helpful. Where possible, constants will be represented by the early letters of the alphabet $a, b, c, \ldots$ while $x, y, z, \ldots$ will be reserved for variables. For random variables we use notations such as $\mathbf{x}$, $\mathbf{y}$, $\mathbf{x}(s)$, and $\mathbf{z}(t)$. Early capital letters will usually be sets, events, or matrices.

We have received much help in writing this material. We appreciate the assistance of the Florida State University secretarial staff, particularly Mrs. Diane Benton and Mrs. Betty Tanner, in preparing the manuscript. We wish to thank Harald Bergström for reading and commenting on an early version, F. W. Leysieffer for many helpful discussions, and the Florida State University statistics faculty for access to their problem collections. We are particularly grateful to R. A. Bradley, "the boss," for the time needed to do the writing. Ingram Olkin has been most helpful in his capacity as consulting technical editor for Holt, Rinehart and Winston, Inc.

W. A. T., Jr.

Columbia, Missouri

May 1969

CONTENTS

1

THE NATURE OF PROBABILITY

1.1 ORIGIN OF PROBABILITY

Gambling is the origin of the mathematical theory of probability. Cardano, who died in 1576, wrote a fifteen-page "gambler's manual" which treated dice and other problems; however, the effective beginning of our subject was in the year 1654. The Chevalier de Méré, in addition to being a knight, was, it appears, somewhat of a gambler. He was concerned over the following problem of "points": A game between two persons is won by the player who first scores three points. Each of the participants places at stake 32 pistoles and the winner takes the entire stake of 64 pistoles. If the two leave off playing when the game is only partly finished, then how should the stakes be divided? For example, if the players have one and two points, respectively, and their chances for winning each point are equal, then what should be the division of stakes? The Chevalier consulted Blaise Pascal who solved the problem and communicated his solution to Fermat. In the ensuing correspondence, the two mathematicians laid a foundation for the theory of probability.

Pascal and Fermat were two of the greatest thinkers of their time. Pascal was a precocious geometer who devoted much of his youth to mathematics and physical science, but he is perhaps best known for his great philosophical classics. Fermat's life was quiet, laborious, and uneventful. He earned his living at a legal profession in the service of France but still found time to pursue pure mathematics as an amusement. The effectiveness with which he pursued his "hobby" can be judged from some of his accomplishments. According to E. T. Bell, he "...conceived and applied the leading idea of the differential calculus thirteen years before Newton was born and seventeen years before Leibniz was born...," and invented analytic geometry inde-

123084

pendent of Descartes but "...his greatest work was the foundation of the theory of numbers." In addition to all of this, Fermat was a cofounder of the theory of probability; he was the "...greatest mathematician of the seventeenth century...."

1.2 THE AXIOMATIC THEORY

High school geometry is perhaps the best known prototype of what mathematics is about. In geometry one starts with undefined concepts such as points and lines, makes assumptions (called axioms) about these concepts, and uses the assumptions to prove theorems. The axiomatic theory of probability, which is the foundation for this book, is mathematics in the same sense. The undefined concepts of probability theory are probability experiment, event, and probability itself.

Standard examples of probability experiments are flipping a coin and rolling a die, but these are misleading because of their symmetry. The experiment of flipping a thumb tack and observing whether it falls point up or point down is a simple intuitive experiment which is not as likely to be misleading. A typical tack will fall "point up" approximately two thirds of the time. See Figure 1.1.

Figure 1.1

The potential outcomes of a probabilistic experiment are called *events*. Some of these events are thought of as being simple or indivisible while others are considered to be compound in that they are composed of simple events connected together by means of the words "and," "or," "not." The simple events have the following two properties: (i) they are exhaustive in the sense that when the experiment is performed one of the simple events always occurs, and (ii) they are mutually exclusive, it being impossible for two simple events to occur on the same performance of the experiment. The illustration of a valve which is to shut off the flow of a liquid at a given instant perhaps provides an example which is more suggestive of the applications. Here the simple events are (i) the flow of liquid is stopped, and (ii) the liquid continues to flow.

Two special events are worth naming. The compound event which contains all outcomes of the experiment is called the *certain event* S and its complement "not S" is called the *impossible event* Φ. It may seem strange to consider the impossible event as a potential outcome of an experiment, but it is convenient to do so much as it is convenient to introduce zero into the number system of arithmetic.

The language and operations of set theory (with which the reader will be familiar) are very useful in discussing compound events, and henceforth, we use these ideas. Table 1.1 will indicate the way in which set operations specialize for events.

Table 1.1 Relation of Set to Event Language

Symbol	*Set Language*	*Event Language*
$A \cap B$	A intersect B	A and B both occur
$A \cup B$	A union B	Either A or B or both occur
$\bar{A}$	Complement of A	A does not occur
S	Universal set	Certain event
Φ	Empty set	Impossible event

If $A \cap B = \Phi$, that is if A and B cannot occur simultaneously, then A and B will be called disjoint sets of mutually exclusive events; in this case, we write $A + B$ instead of $A \cup B$. Unions, intersections, and sums of more than two events will have completely analogous meanings. The notation $A - B$ will signify $A \cap \bar{B}$ where $B \subset A$.

Consider, as an illustration, the probability experiment of rolling a single die and observing the number of dots on the up face. Denoting the simple events by s_1, s_2, s_3, s_4, s_5, and s_6, then $S = s_1 + s_2 + s_3 + s_4 + s_5 + s_6$. The compound event C of observing an even number (that is, 2 *or* 4 *or* 6) is $C = s_2 + s_4 + s_6$. The event $\bar{C}$ of getting an odd number (*not* getting an even number) is $\bar{C} = s_1 + s_3 + s_5$. Of course C and $\bar{C}$ are mutually exclusive compound events since $C \cap \bar{C} = \Phi$.

According to the axiomatic theory, probability is a set or event function; a number is allowed to correspond to each of the events of a probabilistic experiment. The number corresponding to the event C is denoted by $P(C)$ and is called the probability of C. We insist that this correspondence must at least satisfy the following axioms:

Axiom 1

The probability of all events must be non-negative.

Axiom 2

The probability of the certain event is one.

Finite addition axiom

If A and B are mutually exclusive events then $P(A + B) = P(A) + P(B)$.

Several consequences of these axioms are immediately apparent. First, if $A_1, \ldots, A_n$ are finitely many mutually exclusive events then

$$P(A_1 + \cdots + A_n) = P(A_1) + \cdots + P(A_n).$$

Second, for any event C, $P(C) \leq P(C) + P(\overline{C}) = P(S) = 1$ and $P(C)$ is not greater than 1. Third, since S and Φ are disjoint and $S + \Phi = S$ we have $P(S) + P(\Phi) = P(S)$ or $P(\Phi) = 0$.

The purpose of most of the first chapter is to motivate and enlarge on the above choice of axioms. Earlier concepts, such as the equally likely interpretation and frequency probability, led to this choice. The earlier concepts are still important because they provide real life interpretations of abstract mathematical probability, just as light rays and stretched strings are physical realizations of the geometric straight line.

The next three sections discuss three interpretations of the probability concept. An important strength of mathematical disciplines is their capacity to describe many seemingly unrelated aspects of nature. Axiomatic probability is pure mathematics; hence, it will not be surprising to find that the same axioms will be subject to several interpretations.

1.3 THE EQUALLY LIKELY INTERPRETATION OF PROBABILITY

The historically first concept of probability (implicitly utilized by Cardano around 1520) is referred to herein as the equally likely interpretation. This requires that we decide on m equally likely simple events of which c imply the occurrence of a compound event C. The probability of C is then taken to be the ratio of c to m. Thus, if the six faces of a die are taken to be equally likely, and if C denotes the compound event of observing an even number of dots on the up face, then the probability of C, written $P(C)$, is $\frac{3}{6}$.

Note that if a simple events imply A and if A and C are mutually exclusive then $P(C) > 0$, $P(S) = 1$ and

$$P(A + C) = \frac{a + c}{m} = \frac{a}{m} + \frac{c}{m} = P(A) + P(C).$$

Hence the equally likely interpretation satisfies the axioms and is a special case of axiomatic probability.

Further, if c simple events imply C and of these d imply $C \cap B$ then the elementary formula

$$\left(\frac{d}{m}\right) = \left(\frac{c}{m}\right) \cdot \left(\frac{d}{c}\right)$$

yields the important *multiplication rule*:

$$P(C \cap B) = P(C) \cdot P(B|C)$$

where $P(B|C)$ stands for the probability of the event B if the occurrence of C is made part of the conditions of the experiment. $P(B|C)$ is called the *conditional probability* of B given C.

The most important objection to defining probabilities in terms of equally likely events, as is done in some elementary algebra texts, is that many of the most interesting applications cannot be formulated in this way. While the equally likely idea of probability works quite well for rolling a die, in both the valve and the thumb tack examples, it would be exceedingly difficult to intuitively determine m mutually exclusive and equally likely simple events.

Even for the simple experiment of flipping coins all intuitions do not agree on what is equally likely. The mathematician D'Alembert (1754) incorrectly calculated the probability of throwing a head in the course of two throws of a coin. He said that if a head appears on the first throw, the issue is decided, and hence, there are only three cases H, TH, and TT; therefore, the probability is $\frac{2}{3}$. More usual answers are $\frac{2}{4}$ for exactly one head, and $\frac{3}{4}$ for at least one head.

The equally likely concept originates from a time when it was thought necessary to base all mathematics on "self-evident truths," but there is nothing self-evident about equally likely for many applied problems, and hence, it would seem that we must look elsewhere for an adequate concept of probability.

1.4 FREQUENCY PROBABILITIES

If we adopt the point of view that it is essential to be able to verify a probability by observation, and we ask, what is available to check the correctness of a probability, then we are immediately led to the frequency interpretation of probability.

To talk about frequencies, we must first discuss what is meant by n performances of the same experiment. If n experiments differ only in unimportant conditions, then we will say that they are n performances of the same experiment. The apparently necessary vagueness at this point seems to be a drawback to defining probabilities as frequencies. We will see that independence is what is needed here, but this concept will be defined in terms of probability, and hence, a definition of probability in terms of independence would be circular.

If, in n performances of an experiment, a specified event C occurs x times then the *relative frequency* of occurrence of C is x/n.[1] According to the *frequency interpretation*, x/n is a measurement of a permanent numerical property of the experiment; this property is called the frequency probability of C. In short, frequency probability is that property of an experiment which is measured by relative frequency. All measurements are inaccurate to a greater or lesser degree; in the present instance, we would expect the accuracy with which x/n measures $P(C)$ to increase with n. If, in n trials, C occurs x times and the compound event $C \cap B$ occurs x_1 times then $x_1/n = (x/n)\cdot(x_1/x)$ which again yields the multiplication rule:

$$P(C \cap B) = P(C)\cdot P(B|C).$$

There is a difference, however, in that the probabilities involved are now probabilities in the frequency rather than the equally likely sense.

Again, note that relative frequencies satisfy the probability axioms so that frequency probabilities are a realization of axiomatic probabilities.

*1.5 SUBJECTIVE AND ECONOMIC PROBABILITIES

A probabilistic experiment with events $A, B, \ldots$ is to be performed (or has been performed but the result is still unknown). Subjective probability reflects an individual's degree of confidence that some prediction will be verified. This is the meaning of sentences like "The Yankees will probably win the pennant." The time-honored way of measuring an individual's subjective probability is to propose a bet. To place a bet on the event A, a person must find a second individual who is willing to bet on $\bar{A}$, and they must agree on the terms of the bet. A bet on the event A is then a contract which risks losing r dollars if $\bar{A}$ occurs in order to win s dollars if A occurs. This bet will appear *fair* to the person if he would just as soon take one side as the other, that is, if he is indifferent between the two alternatives:

(i) Risking r dollars if $\bar{A}$ occurs to win s dollars if A occurs.
(ii) Risking s dollars if A occurs to win r dollars if $\bar{A}$ occurs.

If a bet of r dollars on A against s dollars on $\bar{A}$ appears fair, then the individual's odds on A are $r:s$ and his (subjective) probability of A is $P(A) = r/(r+s)$.

[1] There is a discrepancy in the literature concerning this terminology. The *Oxford English Dictionary* calls x/n the frequency; we use this word to mean x by itself.

We give an intuitive argument that subjective probabilities satisfy the probability axioms and the multiplication rule. First observe that $P(A) \geq 0$ and $P(A) + P(\bar{A}) = 1$. Concerning the probability of an event S which appears certain we see that an individual should risk an arbitrarily large bet on S in order to win a small amount bet on $\bar{S} = \Phi$. Hence, $P(S) = 1$ and $P(\Phi) = 0$.

To discuss the logical status of subjective conditional probabilities, write $P'(X) = P(X|B)$ for every event X in S. The universal event for the conditional probability experiment has been reduced from S to B but no more information is available about subsets of B. Now, a bet involving two horses should be independent of the other horses in the race. In this way, if X and Y are arbitrary subsets of B, then the ratio of confidence in X to confidence in Y should be unchanged by the knowledge that B has occurred, that is, $P'(X)/P'(Y) = P(X)/P(Y)$, or $P(X)/P'(X) = P(Y)/P'(Y) = k$, a constant. Hence, $P(X) = kP'(X)$ for all subsets X of B. To determine k, note that $P(B) = kP'(B) = k$, and finally, $P(X) = P(B) \cdot P'(X)$ which is the desired result for the special case $X \subset B$. Now, let A be an arbitrary event not necessarily contained in B. However, $A \cap B \subset B$, and hence,

$$\begin{aligned} P(A \cap B) &= P(B) \cdot P'(A \cap B) \\ &= P(B) \cdot P(A \cap B|B) \\ &= P(B) \cdot P(A|B). \end{aligned}$$

Observe that conditional probability is probability in the earlier sense (it reflects an individual's concept of a fair bet), hence, the equation $P(A|B) + P(\bar{A}|B) = 1$ will be valid here as well.

It remains to show that $P(A + B) = P(A) + P(B)$. Letting $A + B = C$, we have then to prove $P(C) = P(A \cap C) + P(\bar{A} \cap C)$. But $P(A \cap C) = P(C)\ P(A|C)$ and $P(\bar{A} \cap C) = P(C)\ P(\bar{A}|C)$. Adding, we obtain

$$P(A \cap C) + P(\bar{A} \cap C) = P(C)[P(A|C) + P(\bar{A}|C)] = P(C) \cdot 1 = P(C).$$

In summary: (i) $P(A) \geq 0$ for all $A \subset S$, (ii) $P(S) = 1$, and (iii) $P(A + B) = P(A) + P(B)$ for disjoint sets A and B. Subjective probabilities satisfy the probability axioms.

Next consider an economic interpretation of probability: "that which is measured by mass betting behavior." The mechanism for betting is as follows: Individual 1 places an agreed-upon amount $x (0 \leq x \leq 1)$ in a pot and receives an A-ticket in return; individual 2 places an amount $1 - x$ in the pot and receives an $\bar{A}$-ticket in return. On this transaction, the price of an A-ticket is x and that of an $\bar{A}$-ticket is $1 - x$. Larger bets are placed by purchasing more tickets. An A-ticket indicates that a pot of size 1 has been set aside until the experiment has been performed and entitles its bearer to receive the entire pot on the occurrence of A.

Denote by $D_A(x)$ the dollar amount of A-tickets which would be purchased if the price were x; $D_A(x)$ is the demand for A-tickets. $D_A(x)$ is arrived at in the following way: The ith individual in the A-ticket market has in mind a personal price a_i (presumably, consistent with his subjective probability of A), which is just barely small enough to induce him to purchase an A-ticket. Here we treat an individual purchasing more than one ticket as several individuals all having the same personal prices. $D_A(x)$ is the number of individuals in the market whose personal prices are not exceeded by x; hence, $D_A(x)$ is a monotone nonincreasing step function.

Similarly $S_A(x)$, the supply of A-tickets at price x, is the dollar amount of A-tickets which would be supplied if the price were x. $S_A(x)$ will be a monotone nondecreasing step function. Note that, because of the mechanism of this market, if an individual demands an A-ticket at price x, then he places an $\bar{A}$-ticket on the market at price $1 - x$; hence, $D_A(x) = S_{\bar{A}}(1 - x)$.

We make the "free market assumption" that the market size is sufficiently great and the spacings of the personal prices are sufficiently close to justify the approximation that the curves $S_A(x)$ and $D_A(x)$ intersect in exactly one point; that is, the equation $S_A(x) = D_A(x)$ has exactly one root which we denote by a. From classical economics, we know that a is the market price of an A-ticket.

Now, let us examine the behavior of the A-ticket prices. First, $a_i \leq 1$, since at best an A-ticket will return 1 dollar; therefore, $D_A(1) = 0$ and $a \leq 1$. On the other hand, $S_A(0) = D_{\bar{A}}(1) = 0$, so that $0 \leq a \leq 1$. Secondly, if a is the price of an A-ticket, that is, $D_A(a) = S_A(a)$, then $D_{\bar{A}}(1 - a) = S_A(a) = D_A(a) = S_{\bar{A}}(1 - a)$, and $\bar{a}$, the price of an $\bar{A}$-ticket, will be $1 - a$; $a + \bar{a} = 1$. Finally, the normal pricing situation, from which there will be some deviation, is that the price of an apple and an orange will be the price of the apple plus the price of the orange. In exactly the same way, the price of an A-ticket and a B-ticket will be their sum whenever $A \cap B = \Phi$. But an A-ticket and a B-ticket are, in effect, exactly the same commodity as an $(A + B)$-ticket, so that their prices should be the same.

Denoting the price of an A-ticket by $P(A)$, we may summarize as follows: (i) $P(A) \geq 0$, (ii) $P(S) = P(A) + P(\bar{A}) = 1$, and (iii)

$$P(A + B) = P(A\text{-ticket and } B\text{-ticket}) = P(A) + P(B)$$

for disjoint sets. The probability axioms provide an appealing model for the ticket prices. We suggest that a probability $P(A)$ may be reasonably interpreted as the free market price of an uncertain prize having value 1 or 0, according as A does or does not occur.

1.6 MORE ON AXIOMATIC PROBABILITY

For axiomatic probabilities, the logical status of the multiplication rule is simply that of a definition. By analogy, with the frequency and equally likely theories, we define *conditional probabilities* so that the multiplication rule will hold. Thus, $P(B|C) = P(B \cap C)/P(C)$, except when $P(C) = 0$, in which case $P(B|C)$ is undefined.

Events B and C are called *independent*, if $P(B \cap C) = P(B) \cdot P(C)$. To attach an intuitive meaning to the concept of independence, we proceed as follows. If B and C are independent then

$$\begin{aligned} P(B \cap \bar{C}) &= P(B) - P(B \cap C) \\ &= P(B)[1 - P(C)] \\ &= P(B) \cdot P(\bar{C}) \end{aligned}$$

so that B and $\bar{C}$ are also independent. Now,

$$P(B|C) = P(B) = P(B|\bar{C});$$

the probability of B is independent (in the grammatical sense) of whether C has or has not occurred.

Three events A, B, and C are said to be independent, if they are independent in pairs and also $P(A \cap B \cap C) = P(A) \cdot P(B) \cdot P(C)$. Similarly n events are independent, if the multiplication formula holds for all choices of n or fewer of the events.

The finite addition axiom of Section 1.2 will suffice for most elementary cases but soon one is confronted with unions of infinitely many events and a satisfactory theory requires

Axiom 3

If $A_1, A_2, \ldots$ are pairwise mutually exclusive events (finite or denumerable in number), then $P(A_1 + A_2 + \cdots) = P(A_1) + P(A_2) + \cdots$

A set function $P(C)$, defined for a class $\mathscr{A}$ of sets closed under countable set operations,[2] and satisfying Axioms 1, 2, and 3, is called a *probability measure*; in a sense, such functions measure the likelihoods with which the sets occur. In their probability context, the sets are called events and Axioms

[2] A class is closed under an operation if performing the operation on things in the class does not yield things outside the class. The motivation for requiring that $\mathscr{A}$ be closed under countable set operations is that we wish to discuss events having structure of this complexity; for example, $A = A_1 + A_2 + \cdots$ and $B = \bigcap_{i=1}^{\infty} B_i$ where A_j and B_i are events. The closure of $\mathscr{A}$ then insures that $P(A)$ and $P(B)$ are defined.

1, 2, and 3 are the *Axioms of Kolmogorov*. Probability measures will form the foundation for all material in this book after Chapter 1.

The properties in the above axioms have been used since the beginning of our subject and many authors have recognized their fundamental role, but Kolmogorov was the first to show, as recently as 1933, that a complete theory of probability could be based on just these three axioms. This demonstration is contained in his monograph *Foundations of Probability* which is still an excellent reference for the mathematically inclined reader.

The importance of the denumerable additivity of probabilities may be seen from its equivalence to the Continuity Property. A probability measure is said to have the *Continuity Property* if

$$\lim_{k \to \infty} P(B_k) = P\left(\bigcap_{k=1}^{\infty} B_k\right),$$

whenever $\{B_k\}$ is a nested decreasing sequence of events.

Theorem 1

In the presence of the finite addition axiom, the continuity property is equivalent to Axiom 3.

Proof

(i) Axiom 3 implies the continuity property. Let $B_1 \supset B_2 \supset \cdots \supset B$, where $B = \bigcap_{k=1}^{\infty} B_k$, then $B_k - B = (B_k - B_{k+1}) + (B_{k+1} - B_{k+2}) + \cdots$, and from Axiom 3

$$P(B_k) - P(B) = P(B_k - B) = P(B_k - B_{k+1}) + P(B_{k+1} - B_{k+2}) + \cdots.$$

In particular, $P(B_1) - P(B) = P(B_1 - B_2) + P(B_2 - B_3) + \cdots$, and the series on the right converges so that its remainder must approach 0; but the remainder after $k - 1$ terms is $P(B_k) - P(B)$. Thus, $\lim_{k \to \infty} P(B_k) = P(B)$.

(ii) Conversely, the Continuity Property implies Axiom 3. For, if A_1, $A_2, \ldots$ are pairwise mutually exclusive events then define $B_k = \sum_{i=k}^{\infty} A_i$. Now, from finite additivity

$$\begin{aligned} P(A_1 + A_2 + \cdots) &= P(B_k) + P(A_1) + \cdots + P(A_{k-1}) \\ &= \lim_{k \to \infty} P(B_k) + P(A_1) + P(A_2) + \cdots \end{aligned}$$

But $\bigcap_{k=1}^{\infty} B_k = \Phi$, for if $s \in \bigcap_{k=1}^{\infty} B_k$, then s is an element of one of the sets $\{A_i\}$, say $s \in A_n$. Then, for $k = 1, 2, \ldots, s \notin A_{n+k}$ because A_n and A_{n+k} are disjoint. Now, $s \notin B_{n+1}$ and $s \notin \bigcap_{k=1}^{\infty} B_k$ which is a contradiction. The

Continuity Property allows us to conclude the proof as follows:

$$\lim_{k \to \infty} P(B_k) = P\left(\bigcap_{k=1}^{\infty} B_k\right) = P(\Phi) = 0.$$

One final elementary but important result will be presented before turning to the applications. Given a partition of the certain event, $S = B_1 + B_2 + \cdots + B_k$, then we may proceed as follows.

$$P(B_1|A) = \frac{P(B_1 \cap A)}{P(A)} = \frac{P(B_1 \cap A)}{\sum_1^k P(B_i \cap A)}$$

$$= \frac{P(A|B_1) \cdot P(B_1)}{\sum_1^k P(A|B_i) \cdot P(B_i)}.$$

This is *Bayes' formula.* A good illustration of its application is provided by the *Problem of the False Positives.* The residents of a community are to be examined for a disease. The examination results are classified as $+$, infection suspected, or as $-$, no indication of infection. But the examination is not infallible. The probability of detecting an infection is only 0.95 and the probability of reporting infection where none exists is 0.01. If 0.2% of the community is diseased, what is the probability of a false $+$? We have

$$P(\text{no infection}|+)$$

$$= \frac{P(+|\text{no infection}) \cdot P(\text{no infection})}{P(+|\text{no infection})P(\text{no infection}) + P(+|\text{infection})P(\text{infection})}$$

$$= \frac{(0.01)(0.998)}{(0.01)(0.998) + (0.95)(0.002)}$$

$$\simeq 0.84.$$

This figure is undesirably high from the medical point of view, but it is unavoidable, since the major proportion of the community is healthy.

1.7 APPLICATION—HEREDITY

Heredity may be defined as the transmission of a quality from parent to offspring. The higher plants and animals are usually formed by the union of two specialized cells called germ cells or gametes. The germ cells unite, grow, and divide, but the resulting cells remain together in an aggregate. Some perform digestive functions, others receive and carry sensation, and still others become specialized as germ cells so that the life cycle may be repeated.

Only the germ cells link successive generations; all others forming the body proper eventually die of old age. In studying heredity, we may think of an organism as having cells of two types: (i) body cells which contain two chromosomes of a kind, and (ii) gametes which contain only one. The chromosomes consist of many small granules connected in sequence by something resembling a thread. These granules may be thought of as containing the hereditary units (genes). The body cells in the same organism contain the same chromosomes, and hence, the same genes. This is so because they are derived from the fertilized egg by a process of successive splitting, each half receiving a complete and identical set of genes. However, cells ear-marked to produce reproductive cells divide in a manner which halves the chromosome number. The number of chromosomes in a body cell is thus constant for all members of a species. In the production of germ cells, the two chromosomes of a kind come together, sometimes exchange parts, and then split,

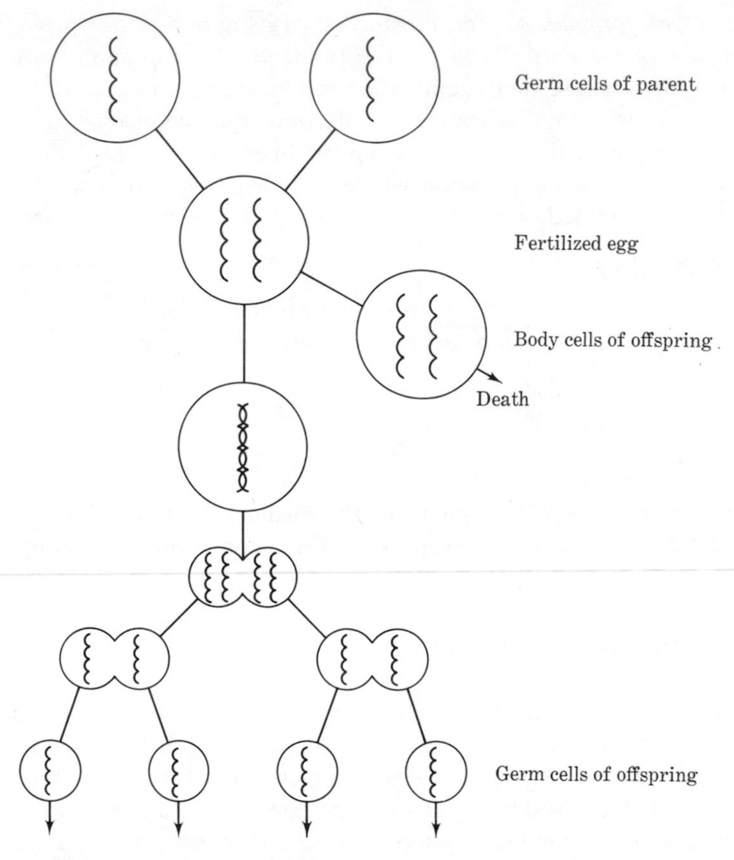

Figure 1.2 Functional Diagram of Multicellular Life Cycle

so that there is a total of four chromosomes of a kind. The cell then splits by successive halving into four parts, each part taking one of the four similar chromosomes with it. This entire process is summarized in Figure 1.2.

Consider, for example, the mating of white, yellow, and cream guinea pigs. A yellow guinea pig has two yellow-color genes in each of its body cells so that only yellow-color genes may be passed on to offspring. Similarly white guinea pigs produce only germ cells having one white-color gene. However, cream guinea pigs have one white- and one yellow-color gene in each body cell, and hence, produce white- as well as yellow-carrying gametes in equal numbers.

If, in mating, the parents are both cream, then we may have white, yellow, or cream offspring. The symmetry of the situation suggests that the following four gamete pairings should be assigned equal probabilities:

$$\text{white}_1 \text{ and white}_2 \qquad \text{white}_1 \text{ and yellow}_2$$
$$\text{yellow}_1 \text{ and white}_2 \qquad \text{yellow}_1 \text{ and yellow}_2$$

where the subscript indicates the parent contributing the gamete. A white offspring can occur in only one of the four equally likely gamete pairings; hence, the probability of a white offspring is taken to be $\frac{1}{4}$. Probabilities determined in this manner are called equally likely probabilities because they are derived by deduction from the symmetric nature of the experiment itself.

1.8 APPLICATION—DOMINANCE OF INHERITED CHARACTERISTICS

In Section 1.7, we showed how the equally likely interpretation of probability may prove to be useful in the study of heredity; there, probabilities were arrived at deductively from the symmetry of the experiment itself. By employing the frequency concept, we may also determine probabilities by induction. This was the method used by Gregor J. Mendel, the Austrian Augustinian monk, who developed much of the theory of Mendelian Inheritance. Mendel's experimental material was the ordinary pea and his work was done in the monastery garden. Among other properties that Mendel studied was the form of the seed. He noticed two different kinds of peas, one round in shape and the other wrinkled. Round and wrinkled peas were crossed and only round peas resulted. This established that the round peas which had been used were purebred, that is, *rr* in genetic makeup. The offspring of this first generation were then allowed to fertilize themselves. The result was 5,474 round peas and 1,850 wrinkled. The relative frequency of wrinkled peas was 1/3.96 or approximately $\frac{1}{4}$. If a great many observations have been recorded, then we may take the relative frequency of an event to be its probability. Accordingly, in the above circumstances we take the frequency probability of a wrinkled pea to be $\frac{1}{4}$. Note that this probability has been arrived at by a process of induction.

Mendel had encountered the property now known as dominance; a pea is round if either of its genes is round, and is wrinkled only if both genes are wrinkled. The genetic reason for Mendel's results is shown in Figure 1.3.

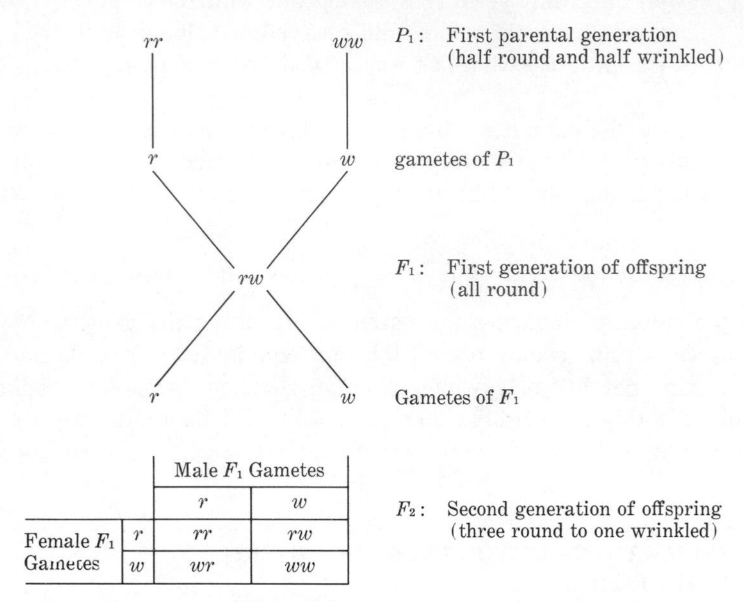

Figure 1.3 Genetic Makeup of Mendel's Experiment

1.9 APPLICATION—LIFE INSURANCE

The amount of the premium is the essential problem which must be solved before life insurance can be sold. What is an appropriate price for insurance on the life of a man, age 25, for a period of one year? A necessary first step here is to determine the probability of death; this is the information contained in a mortality table. An example of a modern mortality table is the Commissioners 1958 Standard Ordinary (Table 1.2). From the table we may read, for example, that the frequency probability of death within the year is 0.00193 for an individual of age 25.

The foundation for all later mortality determinations was laid by John Graunt and Edmund Halley towards the end of the seventeenth century. Graunt analyzed certain London records, called the bills of mortality, which he suggests began to be collected because of the plague. The bills recorded

Table 1.2 Mortality Table (Commissioners 1958 Standard Ordinary)

Age	*Deaths per 1,000*	*Age*	*Deaths per 1,000*	*Age*	*Deaths per 1,000*
0	7.08	34	2.40	67	38.04
1	1.76	35	2.51	68	41.68
2	1.52	36	2.64	69	45.61
3	1.46	37	2.80	70	49.79
4	1.40	38	3.01	71	54.15
5	1.35	39	3.25	72	58.65
6	1.30	40	3.53	73	63.26
7	1.26	41	3.84	74	68.12
8	1.23	42	4.17	75	73.37
9	1.21	43	4.53	76	79.18
10	1.21	44	4.92	77	85.70
11	1.23	45	5.35	78	93.06
12	1.26	46	5.83	79	101.19
13	1.32	47	6.36	80	109.98
14	1.39	48	6.95	81	119.35
15	1.46	49	7.60	82	129.17
16	1.54	50	8.32	83	139.38
17	1.62	51	9.11	84	150.01
18	1.69	52	9.96	85	161.14
19	1.74	53	10.89	86	172.82
20	1.79	54	11.90	87	185.13
21	1.83	55	13.00	88	198.25
22	1.86	56	14.21	89	212.46
23	1.89	57	15.54	90	228.14
24	1.91	58	17.00	91	245.77
25	1.93	59	18.59	92	265.93
26	1.96	60	20.34	93	289.30
27	1.99	61	22.24	94	316.66
28	2.03	62	24.31	95	351.24
29	2.08	63	26.57	96	400.56
30	2.13	64	29.04	97	488.42
31	2.19	65	31.75	98	668.15
32	2.25	66	34.74	99	1,000.00
33	2.32				

christenings and burials along with pertinent information like the sex of a christened child and the cause of death for burials. Graunt published a little book entitled *Natural and Political Observations Made upon the Bills of Mortality*, in which he investigated a number of questions such as the sex ratio of newborn infants, the migration into and out of the city, and the age distribution of the population. Shortly thereafter, Halley prepared the first mortality table from records of births and funerals at the city of Breslau. Halley found the Breslau records more satisfactory for his purpose than Graunt's London tables because Breslau was an isolated city with little immigration or emigration.

1.10 APPLICATION—INFORMATION THEORY

The transmission of information is among the most basic and characteristic of human activities. Before one can compare the efficiencies of alternative information handling systems (verbal, printed, electrical, or other), it is necessary to say quantitatively what is meant by "amount of information." In 1948, Shannon published a mature probabilistic theory which gave a precise meaning to the information concept. Shannon's theory has received a great deal of attention, amounting to general acceptance, in psychology, statistics, probability, and, of course, communications engineering.

Consider a probabilistic experiment which has the simple events $s_1, s_2, \ldots, s_m$, and let $S = \{s_1, s_2, \ldots, s_m\}$. Associate with each of these events a number $p(s_i)$ between 0 and 1 which is the probability that s_i will occur. We say that we have a finite probability experiment

$$\mathbf{S} = \begin{pmatrix} s_1 & s_2 & \cdots & s_m \\ p(s_1) & p(s_2) & \cdots & p(s_m) \end{pmatrix}.$$

In information theory, a typical experiment might consist of observing which punctuation mark or letter of the alphabet occurs at a prescribed place in a given novel. The simple event "a" would have some probability of occurrence and "a" would be more probable than "s."

Before an experiment is performed there is uncertainty about the outcome. The performance of the experiment removes this uncertainty and an individual receives *information* if he is told which of the possible outcomes actually occurred. Information and uncertainty are the same concept looked at from two different points of view and will be taken to be mathematically synonymous. Information theory is an attempt at constructing a model which will make precise the sense in which information has been transmitted to the individual.

It is intuitively clear that of the two alternative experiments

$$\begin{pmatrix} s_1 & s_2 \\ 0.5 & 0.5 \end{pmatrix}, \qquad \begin{pmatrix} s_1 & s_2 \\ 0.99 & 0.01 \end{pmatrix}$$

there is more uncertainty associated with the first since we are almost certain that the outcome of the second experiment will be s_1. Or alternatively, an individual receives more information if he is told the outcome of the first experiment than if he learns the result of the second. Further, more information is received by learning that an experiment has resulted in a rare event than by learning that a very likely occurrence has taken place; it was almost certain that the likely result would occur anyway. Thus, information is inversely associated to the probability of the occurrence of the event. Among the many mathematical forms which will reflect this association, $\log_2 1/p(s) = -\log_2 p(s)$ is defined to be the amount of information conveyed in informing an individual that the event s has occurred. At a later point, we will have more to say concerning this particular choice.

$$H[p(s_1), p(s_2), \ldots, p(s_m)] = -\sum_{S} p(s) \log_2 p(s)$$

is then the average amount of information conveyed in informing an individual of the outcome of the experiment. For this purpose $0 \log 0$ is taken to be 0. We will frequently find it convenient to write $H[p(s_1), p(s_2), \ldots, p(s_m)]$ more briefly as $H(\mathbf{S})$. In accordance with a physical analogy, which need not be discussed here, $H(\mathbf{S})$ is called the *entropy* of the experiment.

Choosing 2 as the base of the logarithm to be used merely fixes the unit of information. If $p(s_1) = p(s_2) = 0.5$, then

$$H(0.5, 0.5) = -0.5 \log_2 0.5 - 0.5 \log_2 0.5 = -\log_2 2^{-1} = 1.$$

This choice is convenient because binary transmission of information is customary (switches on or off, points on a drum magnetized or not, and so forth). The unit of information is called the bit.

PROBLEMS

1. Show that two events are independent and additive if and only if one of them has probability zero.
2. Argue that relative frequencies satisfy the Kolmogorov Axioms.
3. A hunter fires at a rabbit until the rabbit is hit. If the probability of hitting the rabbit on a single round is 0.2, then what is the probability that (a) exactly three rounds will be fired? (b) at least three rounds will be fired?
4. The genetic probability experiment of mating two cream guinea pigs and observing the color of the offspring has eight events which are listed along with their probabilities in the accompanying table. Are there other essentially

different compound events? Do these probabilities satisfy the axioms of Kolmogorov?

Event (color of descendant)	*Probability*
White	$\frac{1}{4}$
Cream	$\frac{1}{2}$
Yellow	$\frac{1}{4}$
White or cream	$\frac{3}{4}$
White or yellow	$\frac{1}{2}$
Cream or yellow	$\frac{3}{4}$
White or cream or yellow	1
Φ	0

5. In a certain animal population the frequency of albino to normal genes is 1:100. Albinoism is recessive (an albino must have two albino genes). Under the assumption of random mating, what will be the proportion of albino offspring in the ensuing generation?

6. The king comes from a family of two children. What is the probability that the other child is his sister?

7. Criticize the following argument: Player A flips a coin and while it is still in the air, B calls heads or tails. But B has a bias; he calls heads 70% of the time. Since heads will occur in only 50% of a large number of flips, B's bias gives A the advantage.

8. If s, t, u, and v are the probabilities that the switches S, T, U, and V will be closed at a given signal, then what is the probability that current flows between the terminals T_1 and T_2 in the following electrical circuit shown in Figure 1.4? What assumptions have you made?

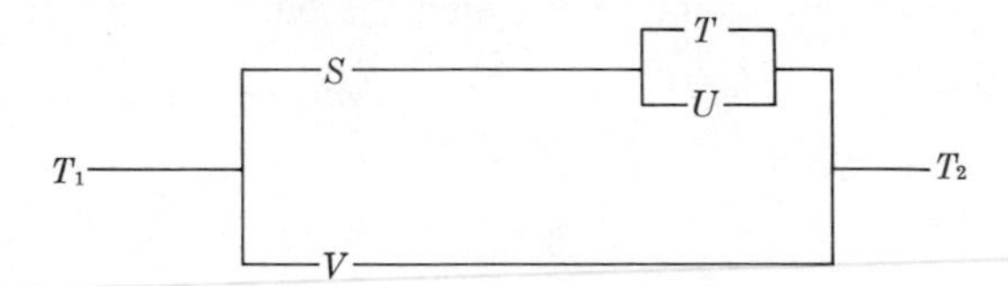

Figure 1.4

9. Prove that if A and B are independent then $\bar{A}$ and $\bar{B}$ are also independent.

10. Construct a theory in terms of physical concepts, such as center of gravity and surface area, which explains the probability experiment of flipping a thumb tack.

11. Comment on the following reasoning of Mairan (1728). A subset of the balls in an urn is chosen at random and a person is asked to guess whether their number n is odd or even. He should guess odd since odd and even are

equally likely if the total number of balls in the urn is even but if the total is odd then the number of cases resulting in n odd will exceed by one the number of cases with n even. For example, if the urn contains three balls then $n = 1$ or 3 are the odd cases and $n = 2$ is the only even case.

12. One purse contains three pennies, a nickel, and two dimes while the other contains five pennies and a dime. A purse is chosen at random and a coin is drawn. What is the probability of drawing a penny?

13. A speaks truth three times out of four, B four times out of five; they agree in asserting that from a bag containing nine balls, all of different colors, a white ball has been drawn; show that the probability that this is true is $\frac{96}{97}$.

14. English and American spellings are *rigour* and *rigor* respectively. A man writes this word and a letter taken at random from his spelling is found to be a vowel. What is the probability that he is an American?

15. Of three cards the first is white on both sides, the second is red on both sides, and the third is white on one side but red on the other. A card is drawn at random and it is found that the up side is white. What is the probability that the down side is also white?

16. A chain is made up of 100 links, assumed to be manufactured independently. The probability that any one breaks when maximum load is applied is 0.001. Find the probability that the chain fails when maximum load is applied.

17. A man owns two old cars, A and B, and has trouble starting them on cold mornings. The probability both will start is 0.1; the probability car B starts and car A does not is 0.2; the probability that neither starts is 0.4.

    (a) Find the probability that car A will start.
    (b) Find the probability that car A will start, given car B starts.
    (c) Find the probability that car B will start given car A starts.

18. Prove that if $P(A \cap B) = P(A \cup B)$, then $P(A) = P(B)$.

19. In a dangerous and complicated engineering system, a safety device is built in. The safety device is supposed to operate if there is a failure in the system. The safety device is designed with the following probabilities: Given the system fails, the probability that the safety device operates is 0.99 and given that the system does not fail, the probability that the safety device does not operate is 0.99. The probability of a failure in the system is assumed to be 0.001.

    (a) Find the probability that the system has failed, given the safety device has operated.
    (b) One Sunday a safety device in the Gemini 6 rocket caused a shutdown 3 seconds after ignition. Comment on this event in relation to the problem and assumptions given for part (a).

20. Show

$$P(A \cup B \cup C) = P(A) + P(B) + P(C) - P(A \cap B) - P(A \cap C) - P(B \cap C) + P(A \cap B \cap C).$$

21. Urn I contains 10 white and 3 red balls. Urn II contains 3 white and 5 red balls. Two balls chosen at random are to be transferred from I to II, and one ball is then to be drawn from II. What is the probability that a white ball will be drawn from II?

2

DISCRETE PROBABILITY MEASURES

2.1 PROBABILITIES DEFINED ON A COUNTABLE SET

The reader may remember from his mathematical studies that a countable set is one which can be put into one-to-one correspondence with a subset of the positive integers. Thus, a countable set is either finite, like the set of faces on a die, or it is denumerable, like the set of odd integers or the set of rationals. The interval [0, 1] is not a countable set of numbers.

Notation such as $\{i\colon s_i \in A\}$, read the set of all i such that s_i is an element of A, will frequently prove useful.

Theorem 2.1

If $S = \{s_1, s_2, \ldots\}$ is a countable set and $\{p_1, p_2, \ldots\}$ is a countable set of nonnegative numbers such that $p_1 + p_2 + \cdots = 1$, then the set function

$$P(A) = \sum_{\{i:\, s_i \in A\}} p_i$$

for all events $A \subset S$, satisfies Kolmogorov's axioms, and hence, is a probability measure.

Proof

The first axiom is satisfied since the p_i's are nonnegative; the second follows from

$$P(S) = \sum_{\{i:\, s_i \in S\}} p_i = 1.$$

Finally

$$P(A + B + \cdots) = \sum_{A+B+\cdots} p_i = \sum_A p_i + \sum_B p_i + \cdots = P(A) + P(B) + \cdots$$

for disjoint sets $A, B, \ldots$. The interchange of the order of summation for the above (potentially) infinite series is justified since $\sum_{A+B+\cdots} p_i$ is absolutely convergent.

Probability measures defined in the manner of the above theorem are called *discrete*.

In Chapter 1, we have shown that impossible events have probability 0. We may ask whether the converse is true. If the numbers $\{p_1, p_2, \ldots\}$ of a discrete probability measure are positive, then $P(A) = 0$ if and only if $A = \Phi$; in fact, if $s_k \in A$, then $P(A) \geq p_k > 0$. But the converse does not hold in general as it is convenient to allow simple events having probability 0. Later in Chapter 4 we will see that, for a general concept of probability, events of probability 0 are more than a convenience, they are needed.

2.2 THE ROLE OF COUNTING IN PROBABILITY

At the very beginning of our discussion of probability, we saw that there are many cases in which the simple events of a probabilistic experiment may be thought of as being equally likely. Thus, in rolling a single die, there will be six basic events, s_1 through s_6, and because of symmetry it will be reasonable to consider these as equally likely; $P(s_i) = \frac{1}{6}$ for $i = 1, 2, \ldots, 6$. More generally, if there are m equally likely simple events, s_1 through s_m, then we may show that the probability of s_i is equal to $1/m$ for $i = 1, \ldots, m$, and if A is any event, then the probability of A equals the number of simple events in the event A divided by m, the total number of simple events. Thus, to evaluate the probability of A, it will be necessary to count or enumerate the number of simple events in A.

With this as motivation, we enter into a review of counting or enumeration in several frequently occurring situations. The following fundamental proposition will be basic to many of our considerations:

Basic Principle

If one operation can be performed in n_1 ways and when it has been performed in any way, a second operation can then be performed in n_2 ways, there will be $n_1 \cdot n_2$ ways of performing the two operations. This result, and its generalization to more than two operations, becomes quite apparent if one considers a tree graph such as that of Figure 2.1.

In enumeration problems, two concepts arise again and again. First, a

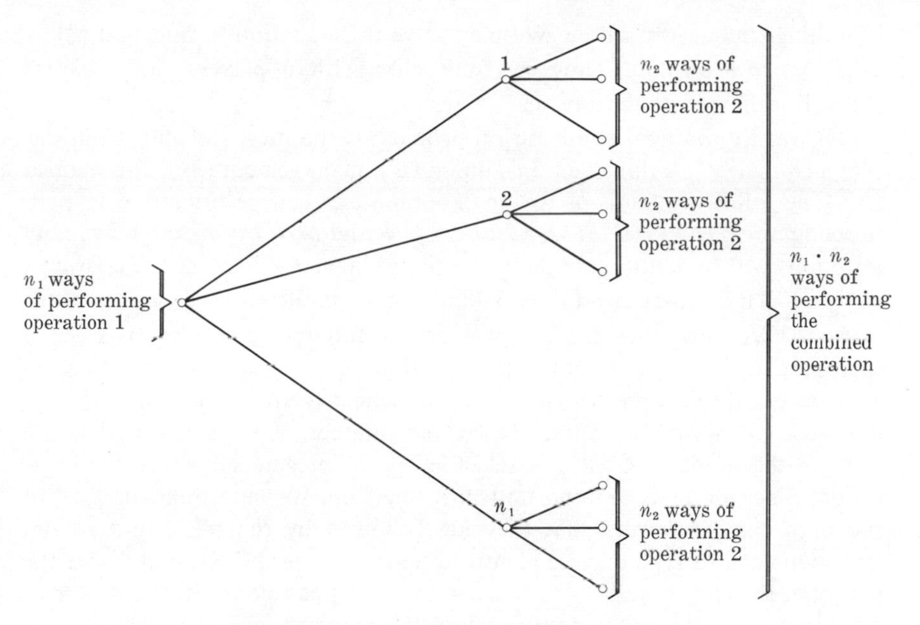

Figure 2.1 The Basic Counting Principle

selection (or combination) of any number of articles means a set of that number classed together but not regarded as having any particular order among themselves. Secondly, an *arrangement* (or permutation) is a set of articles not only classed together, but regarded as having a particular order among themselves.

Four different but related enumerations now arise. First, the number of arrangements of n things r at a time with repetition is n^r. Secondly, the number of arrangements of n things r at a time without repetition, denoted by $n^{(r)}$, is

$$n^{(r)} = n(n - 1)\cdots(n - r + 1). \tag{2.1}$$

These two results follow directly from the basic principle above. Thus, using the letters of the word ABLE, we may form $4^3 = 64$ three-letter words, if letters may be repeated, and $4 \cdot 3 \cdot 2 = 24$ words, if each letter can be used only once. The first result can be used to prove that two people in Columbus, Ohio, have the same initials.

The third problem is the following: in how many ways may we select r out of n physical objects? This would, of course, be selection without repetition. Since every selection of r different objects gives rise to $r!$ different arrangements, we see that $n^{(r)} = r!\binom{n}{r}$, where $\binom{n}{r}$ is the number of selections of n things taken r at a time. We may solve this equation to find that $\binom{n}{r} = n^{(r)}/r! = n!/(n-r)!\,r!$. Thus, a tennis club with 6 players can organize $\binom{6}{2} = 15$ different doubles teams.

Our fourth and final enumeration problem is the most difficult. Consider that a company has three factions, union (U), management (M), and neutral (N). The following kinds of two-man committees can be organized to plan the company picnic: UU, MM, NN, UM, UN, and MN. In general, how many selections with repetition are there of n things taken r at a time? The answer is $\binom{n+r-1}{r}$. Thus, there are $\binom{4}{2} = 6$ kinds of committees made up of three factions. We solve the general problem as follows: Consider that the n things are the first n positive integers. By adjoining $1, 2, \ldots, n$ to each selection of r, we get a selection of $r + n$ integers subject to the condition that each integer appears at least once. Hence, the number of ways required is the same as the number of ways of selecting $n + r$ integers subject to this condition. Since order is unimportant in a selection, we may think that all of the ones come first and that they are followed by the twos, and so on. Selections of this type may be identified by specifying the position where the run of each integer stops. There are $n + r - 1$ possible positions, $n - 1$ of which must be chosen to specify each of the required selections. This can be done in $\binom{n+r-1}{n-1} = \binom{n+r-1}{r}$ ways.

Combinatorial expressions such as $\binom{52}{13}$ are frequently difficult to evaluate numerically. An approximation due to James Stirling is useful for this purpose. *Stirling's formula* is

$$n! \sim (2\pi)^{1/2} n^{n+1/2} e^{-n}$$

in the sense that the ratio of the two sides tends to 1 as $n \to \infty$. Stirling's approximation played an important role in early probability theory.

Three nontrivial illustrations of the use of enumeration in probability now follow.

Example—Probability of a Jack-Queen Contact[1]

What is the probability that at least one Jack and one Queen are adjacent in a shuffled deck of playing cards?

We solve this problem by determining c, the number of selections of 4 Jacks, 4 Queens, and 44 cards for which no Jack is adjacent to a Queen. Then the probability p which we seek is given by

$$p = 1 - \frac{(4!)^2(44!)c}{52!}.$$

[1] Solved in Abad, Jack C., *American Mathematical Monthly*, **72**, 783–784 (1965).

We count these c selections by considering the $\binom{8}{4} = 70$ combinations of 4 Jacks and 4 Queens and dropping cards into appropriate gaps. Thus, the combination J J Q Q Q J J Q has 3 unlike neighbors creating three gaps which must be filled if a Jack is not to be adjacent to a Queen. Let m_i be the number of Jack-Queen combinations having i unlike neighbors. $m_1 = m_7 = 2$, $m_2 = m_6 = 6$, and $m_3 = m_4 = m_5 = 18$. In each combination we drop in i cards, one in each gap between unlike neighbors. The other $44 - i$ cards can be distributed in the 9 positions (front, seven gaps, end) in $\binom{44-i+8}{8}$ ways. So

$$c = \sum_{i=1}^{7} m_i \binom{52 - i}{8} = 27{,}061{,}623{,}270$$

and $p \simeq 0.486$. In 240 performances of this card shuffling experiment, 111 resulted in no Jack-Queen contact. This yields a relative frequency of $\frac{111}{240} = 0.463$ which is fairly close to the theoretical value of 0.486.

Example—The Game of Treize

This problem, first introduced by Montmort around 1714, has maintained a permanent place in our subject. In effect, Montmort asked for the probability p_n that none of n cards, numbered $1, 2, \ldots, n$ and drawn in succession from a bag, will correspond to the order in which it is drawn.

The denominator of p_n will be $n!$ since each of the possible orders is assumed equally likely. A recursion relation for the numerator a_n may be obtained from the following considerations.

In counting a_n, there will be $n - 1$ possible first cards. For each of these, there will be a_{n-2} possible drawings of the form

Order drawn	1	2	3	···	i_1	···	n
Number on card	i_1	i_2	i_3	···	1	···	i_n

with $i_1 \neq 1, i_2 \neq 2, i_3 \neq 3, \ldots, i_n \neq n$, and there will be a_{n-1} possible drawings of the form

Order drawn	1	2	3	···	i_1	···	n
Number on card	i_1	i_2	i_3		k		i_n

$i_1 \neq 1, i_2 \neq 2, i_3 \neq 3, \ldots, k \neq 1, \ldots, i_n \neq n$. Thus

$$a_n = (n - 1)\cdot(a_{n-2} + a_{n-1})$$

and

$$p_n = \frac{a_n}{n!} = \frac{n-1}{n} p_{n-1} + \frac{1}{n} p_{n-2}.$$

From this relation we may establish that

$$p_n = \frac{1}{2} - \frac{1}{3!} + \cdots + \frac{(-1)^n}{n!}$$

and hence $p_n \simeq e^{-1}$ except for n quite small.

In direct generalization of Montmort's question we may ask for $p_n(k)$, the probability that k cards occur in their proper places. We have $p_n(0) = p_n$ and

$$p_n(k) = \frac{\binom{n}{k}a_{n-k}}{n!} = \frac{p_{n-k}}{k!}$$

$$\simeq \frac{e^{-1}}{k!}$$

for fixed k and large n. In Section 3.3 we will discuss the Poisson probability function of which this approximation is a special case.

Pierre Remond de Montmort (1678–1719) led a simple retired life devoted to religion, philosophy, and mathematics. His work *Essai d'Analyse sur les Jeux de Hazard* is divided into four parts treating (i) combinations, (ii) card games, (iii) dice games, and (iv) problems, including five famous problems proposed by Huygens, which are presented at the end of this chapter.

Example—Use of the Reflection Principle

Balls are drawn successively from a box containing m white and n black balls ($m > n$). What is the probability that at some time the number of white and black balls drawn will be equal?

On the grid of Figure 2.2 plot the number of white balls on the vertical axis and the number of black on the horizontal. The result of drawing $m + n$ balls is a path from (0, 0) to (n, m). These $\binom{m+n}{m}$ paths are equally likely (verify this).

The desired probability will be $p = c/\binom{m+n}{m}$ where c is the number of paths from (0, 0) to (n, m) which touch the diagonal line joining (0, 0) to (n, n). Such paths are of two types (i) those visiting the point (1, 0), all of which intersect the diagonal line, and (ii) those which visit the point (0, 1) and then touch the diagonal. There is a one-to-one correspondence between the two types of paths and hence they are equally numerous. This correspondence, called the reflection principle, is illustrated in Figure 2.2. There are $\binom{m+n-1}{m}$ paths of type (i); thus,

$$p = \frac{2\binom{m+n-1}{m}}{\binom{m+n}{n}} = \frac{2n}{m+n}.$$

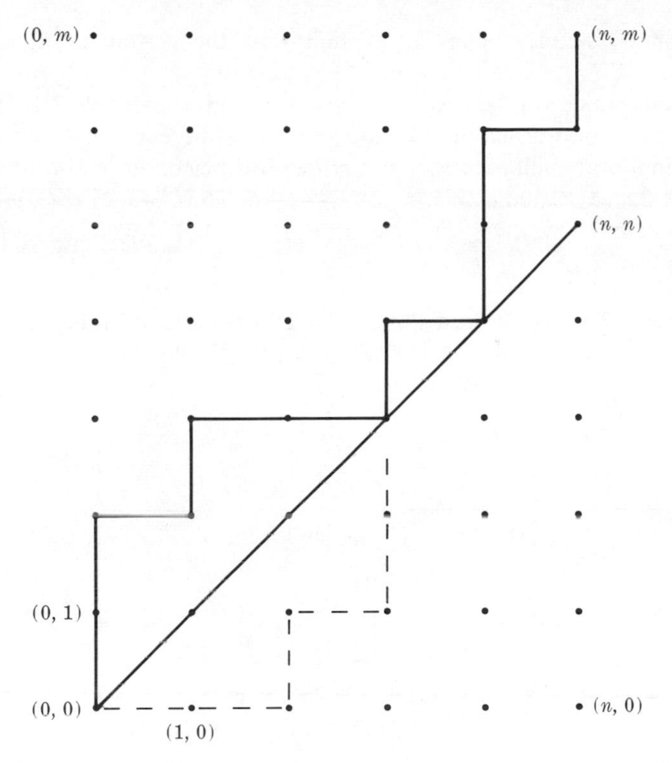

Figure 2.2 The Reflection Principle

2.3 JOINT, MARGINAL, AND CONDITIONAL PROBABILITIES

Frequently we will want to consider the joint performance of two or more probability experiments. Thus, we might wish to observe the sum as well as the difference of the number of dots on the up face of a red and green die. If the 36 potential results of rolling a red and a green die are taken to be equally likely, then (from the axioms) we obtain the joint probabilities of Table 2.1. The probabilities of the various sums and differences are (according to the axioms) calculated by totaling the rows and columns respectively; hence these latter probabilities are called *marginal probabilities*.

According to the multiplication rule, conditional probabilities will be calculated by dividing a joint by a marginal probability. For example, the conditional probability of a sum of 8 given a difference of 2 is $(\frac{1}{36})/(\frac{4}{36}) = \frac{1}{4}$. Marginal and conditional probabilities do not differ in any essential way from ordinary probabilities; the word marginal emphasizes that there is at least one other experiment which is simultaneously being considered while "conditional" simply makes it explicit that the certain event has been restricted.

Notations which will be used may be illustrated from Table 2.1. We write $p(s, d)$ for the joint probability of the sum and difference, $p(s)$ and $p(d)$ for the marginal probabilities, $p(s|d)$ and $p(d|s)$ for the conditional probabilities. Here, $s = 2, \ldots, 12$ and $d = 5, 4, \ldots, -4, -5$. Thus, $p(3, 2) = 0$, $p(8) = \frac{5}{36}$ for the sum, and $p(8|2) = \frac{1}{4}$ for the conditional probability of a sum of 8 given a difference of 2.

Table 2.1 The Tabled Entry is 36 Times the Joint Probability of a Sum and Difference when Rolling Two Dice

Sum / *Difference (red-green)*	2	3	4	5	6	7	8	9	10	11	12	*Marginal probabilities of difference (times 36)*
5						1						1
4					1		1					2
3				1		1		1				3
2			1		1		1		1			4
1		1		1		1		1		1		5
0	1		1		1		1		1		1	6
−1		1		1		1		1		1		5
−2			1		1		1		1			4
−3				1		1		1				3
−4					1		1					2
−5						1						1
Marginal probabilities of sum (times 36)	1	2	3	4	5	6	5	4	3	2	1	

Perhaps the simplest way in which experiments may be related is if they are independent. We have previously defined independent events; now generalizing, two experiments are said to be *independent* if $P(C \cap B) = P(C) \cdot P(B)$ whenever C and B are possible results of the first and second experiments, respectively. From the multiplication rule we see that this is the same as requiring $P(B) = P(B|C)$, that is, the probability of every possible result of the second experiment is independent in the grammatical sense of what has happened on the first experiment. Note that the two highly related experiments of Table 2.1 do not satisfy the definition of independence.

If the potential results and their probabilities are identical for a sequence of n independent experiments, then the sequence is said to consist of n independent trials. Earlier when we introduced the frequency definition of probability, it was the concept of independent trials which was needed but unavailable. Of course, the idea of independent trials could be developed in the above manner from either the equally likely or the frequency definitions, but independence would be available only after probability is introduced, and hence, could not be made the basis for defining probability. Observing the result of tossing a thumb tack n times is an elementary example of a sequence of independent trials. Each trial has two potential outcomes which we may call "up" and "down"; the probability of "up" is the same on each trial. This kind of sequence is important enough to name. In general, we will say that we have a sequence of n *Bernoulli trials if*:

(i) Each trial has two possible outcomes which we may call "success" and "failure."
(ii) The probability of "success" is the same for every trial.
(iii) The n trials are independent probability experiments.

It is customary to write p for the probability of "success" on a single trial. The probability of a failure is then $1 - p$ ($= q$ say). The 2^n potential results of a sequence of n Bernoulli trials may be represented by ordered sequences of S's and F's. As an example, we might wish to write

$$P(SFFFS) = p \cdot q \cdot q \cdot q \cdot p = p^2 q^3.$$

(Give step by step reasons why this is correct.)

Bernoulli trials are named for James Bernoulli, one member of a large family who collectively played a leading role in developing the calculus and its applications during the seventeenth and eighteenth centuries. The Bernoullis were one of many Protestant families who fled from Antwerp in 1583 to escape massacre by the Catholics; they settled at Basle, Switzerland. James, Nicolas, and Daniel Bernoulli all made major contributions to Probability. James' work is collected in his book, the *Ars Conjectandi*, which was posthumously published in 1713.

2.4 APPLICATION—COMMUNICATION CHANNELS

This and the next two sections are continuations of the information theory example begun in Section 1.10. A communication channel may be thought of as consisting of two probability experiments (1) the transmitted signal with possible outcomes $S = \{s_1, s_2, \ldots, s_m\}$ and (2) the received signal with possible outcomes $T = \{t_1, t_2, \ldots, t_\ell\}$. (1) and (2) differ from each other because of "noise" which is present in the transmission line. This may be represented schematically as in Figure 2.3.

What is meant by the capacity of a channel to transmit information? We motivate a definition as follows. For each possible received signal t, there is uncertainty concerning the signal s which was sent. This uncertainty is

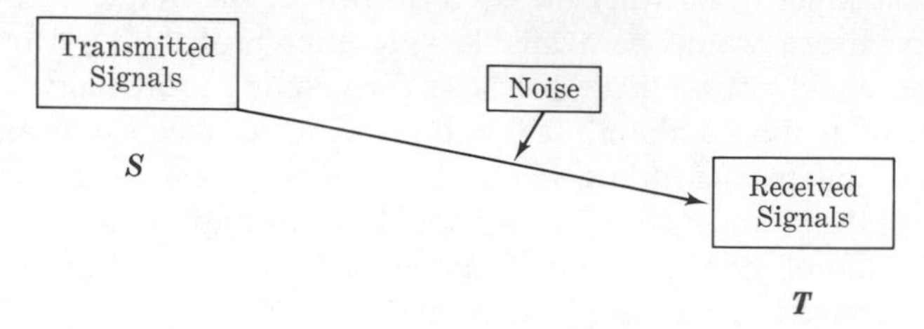

Figure 2.3 Scheme of Communication Channel

measured by $H(\mathbf{S}|t) = -\sum_S p(s|t) \log p(s|t)$. Averaging over all possible received signals gives

$$H(\mathbf{S}|\mathbf{T}) = \sum_T p(t)H(\mathbf{S}|t) = -\sum_{S,T} p(s, t) \log p(s|t)$$

which is the average uncertainty about $\mathbf{S}$ given $\mathbf{T}$. Subtracting from $H(\mathbf{S})$, the total uncertainty about $\mathbf{S}$, we obtain the average uncertainty about $\mathbf{S}$ which has been removed by observing $\mathbf{T}$. Finally then, the *capacity c of a channel* to remove uncertainty (or transmit information) is

$$c = \max_{p(s)} [H(\mathbf{S}) - H(\mathbf{S}|\mathbf{T})].$$

Intuitively, we would want the capacity of a perfect channel (one without noise) to be the maximum information of the transmitted signal. Let us see if this is true with the above definition. For a perfect channel, we would have

$$p(s|t) = \frac{p(s, t)}{p(t)} = \begin{cases} 0 & s \neq t \\ 1 & s = t \end{cases}$$

so that $H(\mathbf{S}|\mathbf{T}) = H(\mathbf{S}|t) = 0$. As expected

$$c = \max_{p(s)} H(\mathbf{S}) = \max_{p(s)} \left\{ -\sum_S p(s) \log p(s) \right\}.$$

Later, in Theorem 2.3, we will evaluate this maximum explicitly.

In the special case of a *binary symmetric channel*, it is possible to send only two messages which we may denote by 0 and 1. The probabilities of correct and incorrect transmission are p and q, respectively, regardless of the message being sent. This is summarized in Table 2.2

Table 2.2 $p(s|t)$ for Binary Symmetric Channel

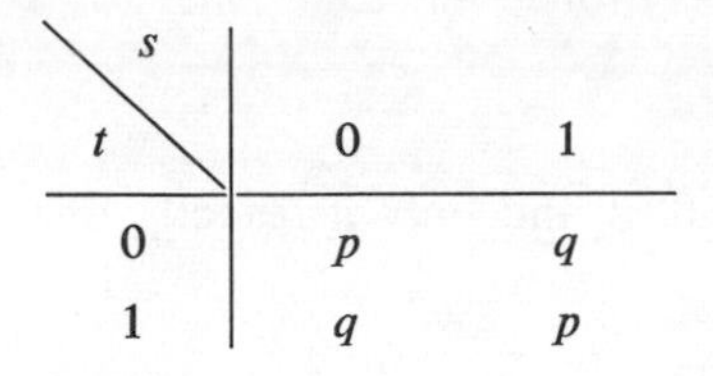

$t \backslash s$	0	1
0	p	q
1	q	p

$$H(\mathbf{S}|\mathbf{T}) = H(\mathbf{S}|t = 0) = H(\mathbf{S}|t = 1) = -p \log p - q \log q = H^*, \text{ say.}$$

Now introducing the notation $p(s_0) = x$ and $p(s_1) = (1 - x)$ then

$$H(\mathbf{S}) = -x \log x - (1 - x) \log (1 - x).$$

Thus,

$$H(\mathbf{S}) - H(\mathbf{S}|\mathbf{T}) = -x \log x - (1 - x) \log (1 - x) - H^* = f(x), \text{ say.}$$

The capacity of the channel is then calculated by differentiating $f(x)$ and equating to zero to find that $f(x)$ is maximized at $x = 0.5$. Thus

$$c = \max_x f(x) = f(0.5) = 1 + p \log p + q \log q.$$

A graph of c appears in Figure 2.4.

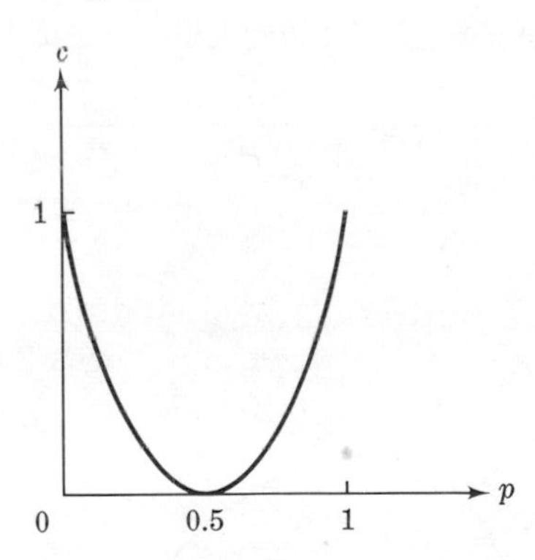

Figure 2.4 Capacity of the Binary Symmetric Channel

For $p = \frac{1}{2}$, the capacity is 0 since the received signal gives no clue concerning the transmitted signal. At the other extreme for a channel without noise, the capacity is 1, the maximum uncertainty of a two-state experiment.

2.5 APPLICATION—PROPERTIES OF ENTROPY

In the interest of simplifying the notation, it will sometimes be advisable to write p_i instead of $p(s_i)$ for $i = 1, 2, \ldots, m$.

Theorem 2.2

$H(p_1, p_2, \ldots, p_m) = 0$, if and only if, exactly one of the numbers $p_1, p_2, \ldots, p_m$ equals 1.

Proof

$H \geq -p_k \log p_k \geq 0$, for $k = 1, 2, \ldots, m$ and $x \log x = 0$ only if $x = 0$ or $x = 1$. Thus, if $H = 0$, then $p_k = 0$, or $p_k = 1$ for $k = 1, 2, \ldots, m$. But $\sum_i p_i = 1$, and hence, not all p_k can be zero. Conversely, if one of the p's (say p_1 for definiteness) equals 1 then the remainder must be zero and $H = 0$.

The reader will remember that a concave function is defined by the property that its path lies above every chord drawn to it; or analytically, $f(x)$ is concave if $\lambda_1 f(x_1) + \lambda_2 f(x_2) \leq f(\lambda_1 x_1 + \lambda_2 x_2)$, whenever $\lambda_1, \lambda_2 \geq 0$ and $\lambda_1 + \lambda_2 = 1$. It is clearly equivalent to require that a convex function satisfy $\sum_k \lambda_k f(x_k) \leq f(\sum_k \lambda_k x_k)$, whenever $\lambda_k \geq 0$ and $\sum_k \lambda_k = 1$.

From Figure 2.5, we see that $f(x) = -x \log_2 x$ is a continuous and concave function.

Theorem 2.3

$$H(p_1, p_2, \ldots, p_m) \leq H(1/m, 1/m, \ldots, 1/m) = \log_2 m.$$

x	0	1/4	$e^{-1} = 0.367$	1/2	3/4	1
$-x \log_2 x$	0	1/2	0.53	1/2	0.31	0

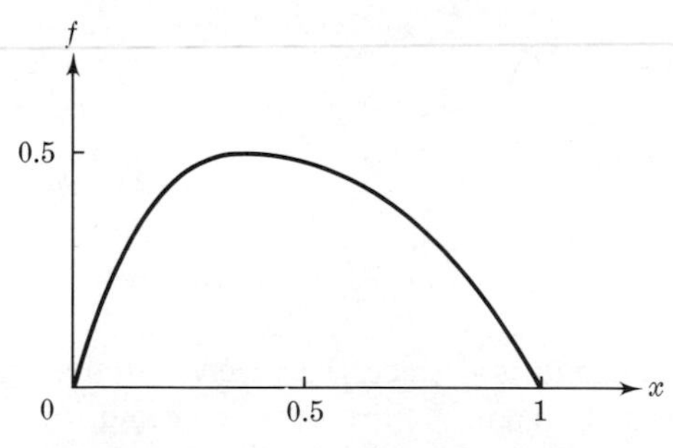

Figure 2.5 Graph of $f(x) = -x \log_2 x$

Proof

If we write

$$f(x) = -x \log x, \lambda_1 = \lambda_2 = \cdots = \lambda_m = \frac{1}{m}$$

and

$$x_i = p_i \quad \text{for } i = 1, 2, \ldots, m,$$

then the concave function property specializes to

$$\sum_k \frac{1}{m}(-p_k \log p_k) \leq -\left(\sum_{k=1}^{m} \frac{1}{m} p_k\right) \log\left(\sum_k \frac{1}{m} p_k\right).$$

Since $\sum_k p_k = 1$, this expression can be simplified to $-\sum_k p_k \log p_k \leq -\log(1/m)$ which completes the proof.

Theorem 2.4

$H(\mathbf{S}, \mathbf{T}) = H(\mathbf{T}) + H(\mathbf{S}|\mathbf{T})$.

Proof

$p(s, t) = p(t)p(s|t)$.

Thus

$$\begin{aligned} H(\mathbf{S}, \mathbf{T}) &= -\sum_{S,T} p(s, t)[\log p(t) + \log p(s|t)] \\ &= -\sum_{T} \log p(t) \sum_{S} p(s, t) - \sum_{S,T} p(s, t) \log p(s|t) \\ &= H(\mathbf{T}) + H(\mathbf{S}|\mathbf{T}). \end{aligned}$$

From the second previous line, we see that if **S** and **T** are independent experiments, that is, $p(s|t) = p(s)$, then

$$H(\mathbf{S}, \mathbf{T}) = H(\mathbf{T}) - \sum_{S,T} p(s, t) \log p(s) = H(\mathbf{T}) + H(\mathbf{S}).$$

That is, the informations of two independent experiments add to give the information of the compound experiment. On the other hand, if the result of **T** uniquely determines the result of **S** then for each t, $p(s|t)$ equals 1 or 0 for all s in S. Hence for each s and t,

$$p(s, t) \log p(s|t) = p(t)p(s|t) \log p(s|t) = 0, \quad H(\mathbf{S}|\mathbf{T}) = 0$$

and

$$H(\mathbf{S}, \mathbf{T}) = H(\mathbf{T}).$$

This agrees with intuition since it says that if the outcome of **S** is determined by the outcome of **T** then the average information of **T** is the average information of the compound experiment (**S**, **T**).

Theorem 2.5

$H(\mathbf{S}|\mathbf{T}) \leq H(\mathbf{S})$.

This says that the uncertainty about an experiment **S** can only be decreased by knowing the outcome of an experiment **T**. Theorem 2.5 also has an important implication for communication theory; it says that the capacity of a channel must be nonnegative.

Proof

In the concave function property take $f(x) = -x \log x$, $\lambda_k = p(t_k)$ and $x_k = p(s|t_k)$. Then

$$\sum_k \lambda_k x_k = \sum_T p(t)p(s|t) = \sum_T p(s, t) = p(s)$$

and

$$\sum_T p(t)[-p(s|t) \log p(s|t)] \leq -p(s) \log p(s).$$

This expression may be simplified to read

$$-\sum_T p(s, t) \log p(s|t) \leq -p(s) \log p(s)$$

and the theorem is proved by summing over S.

*2.6 APPLICATION—THE UNIQUENESS THEOREM OF INFORMATION THEORY

Among the properties which entropy "should" have, the following three seem to be essential.

(i) $H(p_1, p_2, \ldots, p_m) \leq H(1/m, 1/m, \ldots, 1/m)$,
(ii) $H(\mathbf{S},\mathbf{T}) = H(\mathbf{T}) + H(\mathbf{S}|\mathbf{T})$, where $H(\mathbf{S}|\mathbf{T}) = \sum_T p(t)H(\mathbf{S}|t)$,
(iii) $H(p_1, p_2, \ldots, p_m, 0) = H(p_1, p_2, \ldots, p_m)$.

Theorem 2.6

If H is a continuous function satisfying the above three properties and defined for each positive integer m and for all nonnegative $p_1, p_2, \ldots, p_m$

such that $\sum_{i=1}^{m} p_i = 1$, then

$$H(p_1, p_2, \ldots, p_m) = -\lambda \sum_{i=1}^{m} p_i \log p_i,$$

where λ is a nonnegative constant.

In particular, the above theorem states that

$$L(m) \equiv H\left(\frac{1}{m}, \frac{1}{m}, \ldots, \frac{1}{m}\right) = -\lambda \sum_{i=1}^{m} \frac{1}{m} \log m^{-1} = \lambda \log m.$$

We prove this special case first.

Lemma

Under the conditions of the theorem, $L(m) = \lambda \log m$, where $\lambda > 0$.

Proof

$$L(m) = H\left(\frac{1}{m}, \frac{1}{m}, \ldots, \frac{1}{m}, 0\right) \leq H\left(\frac{1}{m+1}, \frac{1}{m+1}, \ldots, \frac{1}{m+1}\right) = L(m+1).$$

Let

$$\mathbf{S}_i = \begin{pmatrix} s_{1i} & s_{2i} & \cdots & s_{ri} \\ 1/r & 1/r & \cdots & 1/r \end{pmatrix}$$

for $i = 1, 2, \ldots, \ell$, be ℓ independent experiments.

$$\begin{aligned} L(r^\ell) &= H(\mathbf{S}_1\mathbf{S}_2 \cdots \mathbf{S}_\ell) = H(\mathbf{S}_1) + H(\mathbf{S}_2) + \cdots + H(\mathbf{S}_\ell) \\ &= \ell L(r). \end{aligned}$$

Likewise $L(t^k) = kL(t)$ for integral k and t. Let r, ℓ, and t be arbitrary positive integers with $t \neq 1$, but choose k so that

$$t^k \leq r^\ell \leq t^{k+1}.$$

$k \log t \leq \ell \log r \leq (k+1) \log t$ and

$$\frac{k}{\ell} \leq \frac{\log r}{\log t} \leq \frac{k}{\ell} + \frac{1}{\ell}.$$

Because of the monotone nature of L, $L(t^k) \leq L(r^\ell) \leq L(t^{k+1})$ or $kL(t) \leq \ell L(r) \leq (k+1)L(t)$. Hence,

$$\frac{k}{\ell} \leq \frac{L(r)}{L(t)} \leq \frac{k}{\ell} + \frac{1}{\ell}$$

and

$$\left|\frac{L(r)}{L(t)} - \frac{\log r}{\log t}\right| \leq \frac{1}{\ell}.$$

Thus, since ℓ is arbitrary,

$$\frac{L(r)}{L(t)} = \frac{\log r}{\log t}$$

and letting $t = 13$, say

$$L(r) = \frac{L(13)}{\log 13} \cdot \log r = \lambda \log r.$$

Since $L(r)$ is nondecreasing $\lambda \geq 0$. This concludes the proof of the lemma.

Now consider an experiment

$$\mathbf{A} = \begin{pmatrix} a_1 & a_2 & \cdots & a_m \\ p_1 & p_2 & \cdots & p_m \end{pmatrix}$$

where all p_i are rational. Since $\sum_{i=1}^{m} p_i = 1$ we may write $p_i = g_i/g$, where all g_i are positive integers and $g = \sum_{i=1}^{m} g_i$. Consider also a joint experiment **AB** with joint probabilities given by Table 2.3.

Table 2.3 Joint Probabilities for the Experiment **AB**

	b_{11}	b_{12}	$\cdots$	b_{1g_1}	b_{21}	b_{22}	$\cdots$	b_{2g_2}	$\cdots$	b_{m1}	b_{m2}	$\cdots$	b_{mg_m}	$p(a_i)$
a_1	g^{-1}	g^{-1}	$\cdots$	g^{-1}	0	0	$\cdots$	0	$\cdots$	0	0	$\cdots$	0	g_1/g
a_2	0	0	$\cdots$	0	g^{-1}	g^{-1}		g^{-1}	$\cdots$	0	0	$\cdots$	0	g_2/g
$\vdots$	$\vdots$	$\vdots$		$\vdots$	$\vdots$	$\vdots$		$\vdots$		$\vdots$	$\vdots$		$\vdots$	$\vdots$
a_m	0	0	$\cdots$	0	0	0	$\cdots$	0	$\cdots$	g^{-1}	g^{-1}	$\cdots$	g^{-1}	g_m/g

Notice that the marginal probabilities for the experiment A are as given above. Since b_{ij} can occur only if a_i has occurred, the experiment **AB** has g possible outcomes and $P(\{a_i\} \cap \{b_{ij}\}) = g^{-1}$ for all i and j. Thus, $H(\mathbf{AB}) = L(g) = \lambda \log g$ according to the lemma. Also, $H(\mathbf{B}|a_i) = \lambda \log g_i$ according to the lemma.

$$\begin{aligned} \lambda \log g = H(\mathbf{AB}) &= H(\mathbf{A}) + H(\mathbf{B}|\mathbf{A}) \\ &= H(\mathbf{A}) + \sum_{i=1}^{m} p_i H(\mathbf{B}|a_i) = H(\mathbf{A}) + \sum_{i=1}^{m} p_i \lambda \log g_i. \end{aligned}$$

$$\lambda \log g = H(\mathbf{A}) + \lambda \sum p_i \log p_i + \lambda \log g,$$

and finally

$$H(\mathbf{A}) = -\lambda \sum_{i=1}^{m} p_i \log p_i,$$

which proves the theorem for rational p_i. From the continuity of H, it is now clear that the theorem must be valid for arbitrary p_i as well. The theorem is proved.

PROBLEMS

The following 5 problems played an important role in the early development of probability; they were posed without analysis or demonstration by Huygens in a treatise entitled *De Ratiocinii in Ludo Aleoe* around 1657. Solutions were given by Bernoulli in his probability text, the *Ars Conjectandi*. We quote directly from Todhunter (1865), *A History of the Mathematical Theory of Probability*.

1. A and B play with two dice on this condition, that A gains if he throws six, and B gains if he throws seven. A first has one throw, then B has two throws, then A has two throws, and so on until one or the other gains. Show that A's chance is to B's as 10,335 to 12,276.

2. Three players A, B, C take twelve balls, eight of which are black and four white. They play on the following conditions; they are to draw blindfold, and the first who draws a white ball wins. A is to have the first turn, B the next, C the next, and A again, and so on. Determine the chances of the players.

Bernoulli solves this on three suppositions as to the meaning; first he supposes that each ball is replaced after it is drawn; secondly he supposes that there is only one set of twelve balls, and that the balls are not replaced after being drawn; thirdly he supposes that each player has his own set of twelve balls, and that the balls are not replaced after being drawn.

3. There are forty cards forming four sets each of ten cards; A plays with B and undertakes in drawing four cards to obtain one of each set. Show that A's chance is to B's as 1,000 is to 8,139.

4. Twelve balls are taken, eight of which are black and four are white. A plays with B and undertakes in drawing seven balls blindfold to obtain three white balls. Compare the chances of A and B.

5. A and B each take twelve counters and play with three dice on this condition, that if eleven is thrown A gives a counter to B, and if fourteen is thrown B gives a counter to A; and he wins the game who first obtains all the counters. Show that A's chance is to B's as 244,140,625 is to 282,429,536,481.

6. Three cards are drawn at random from a deck. What is the probability that they are a Jack, Queen, and King?

7. Verify that Figure 2.5 is, in fact, a graph of the function $f(x) = -x \log_2 x$.

8. How much information is transmitted in performing n Bernoulli trials?

9. If **S** and **T** are the experiments of noting the sum and the difference respectively in the two-dice example of Table 2.1, then calculate $H(\mathbf{S}|\mathbf{T})$.

10. Twenty-three people take their seats at a round table; show that the odds are ten to one against two specified people sitting together.

11. The four letters s, e, n, t are randomly arranged in a row. What is the probability that they form an English word?

12. In a bag there are five white and four black balls. If they are withdrawn one at a time what is the probability that the first will be white, the second black, and so on alternately?

13. If a letter is taken at random from the word *organize*, what is the probability that it is a vowel?

14. If two letters are taken at random from *murmurer*, what is the probability that they are alike?

15. A letter is chosen at random out of each of the words *musical* and *amusing*; what is the probability that the same letter is chosen in each case?

16. Three different persons each name an integer not greater than n. Find the probability that the integers named are such that the sum of every two is greater than the third.

17. Consider a bridge hand in which there are no jacks, queens, kings, or aces. What is the probability of such a hand being dealt?

18. Consider a bridge hand in which there are no aces. West reports that in six successive hands, exactly five were of this type. What is the probability that this should occur?

19. A drawer contains four black, six brown, and two blue socks. Two socks are taken at random from the drawer, one after the other.
    (a) What is the probability that both socks will be of the same color?
    (b) Given that the two socks are of the same color, what is the probability that they are blue?
    (c) Show that the event: "Second sock is blue," and the event: "First sock is black," are dependent events.

3

DISCRETE RANDOM VARIABLES

3.1 RANDOM VARIABLES, PROBABILITY FUNCTIONS, AND EXPECTATION

A *random variable* (abbreviated r.v.) $\mathbf{x}(s)$ is a function defined for each of the simple events of some probability experiment and having the real line as range[1]; $\mathbf{x}(s)$ is *discrete* if its range is some countable subset $X = \{x_1, x_2, \ldots\}$ of the real line. That is, to each simple event $s \in S$, there corresponds exactly one of the numbers $x_1, x_2, \ldots$; if x corresponds to s, then we write $\mathbf{x}(s) = x$. The probability $P_{\mathbf{x}}(A)$, that $\mathbf{x}$ is an element of some set A in the range of $\mathbf{x}$ is the probability of the set of all events in S which map into A, that is, $P_{\mathbf{x}}(A) = P[\{s \in S : \mathbf{x}(s) \in A\}]$. The function $p_{\mathbf{x}}(x) = P(\mathbf{x}(s) = x)$ will be called the *probability function*[2] of $\mathbf{x}$. Frequently, we will write $\mathbf{x}$ for $\mathbf{x}(s)$ and $p(x)$, instead of $p_{\mathbf{x}}(x)$; the reader is then expected to fill in these omissions from the context.

As a first simple example, consider the game of matching pennies. If Peter matches Paul then Peter wins a cent and otherwise Paul wins one penny. (These two seem to play a major role in the literature of our subject.) Let $\mathbf{x}(s)$ be "Peter's winnings." The simple events may be identified by a notation such as HT, where the result of Peter's coin (a head in this case) appears first. We have $\mathbf{x}(HH) = \mathbf{x}(TT) = +1$ and $\mathbf{x}(HT) = \mathbf{x}(TH) = -1$ while $p_{\mathbf{x}}(+1) = P(\mathbf{x} = +1) = \frac{1}{2}$ and $P(\mathbf{x} = -1) = p_{\mathbf{x}}(-1) = \frac{1}{2}$.

If Peter and Paul play this game n times and if x matches result, then Peter's average winnings will be

$$\frac{x - (n - x)}{n} = 1 \cdot \frac{x}{n} + (-1)\left(1 - \frac{x}{n}\right).$$

[1] Actually, an additional condition of measurability is required; see Chapter 8.

[2] This terminology is not usual but we find it convenient.

NORTHWEST MISSOURI
STATE COLLEGE LIBRARY
MARYVILLE, MISSOURI

But x/n is a measure of the probability of a match, and hence, Peter's average winnings will be a measure of $1 \cdot P(\mathbf{x} = 1) + (-1)P(\mathbf{x} = -1)$. We describe this situation by saying that Peter will "expect" to win

$$1 \cdot P(\mathbf{x} = 1) + (-1)P(\mathbf{x} = -1) = 0$$

(cents) each time he matches coins with Paul. Thus, on the average Peter should "expect" neither to win nor lose in a long sequence of matches. More generally, we define the *expectation* of any discrete random variable $\mathbf{x}(s)$ as follows:

$$E\mathbf{x}(s) = \sum_{i=1}^{\infty} x_i \cdot P[\mathbf{x}(s) = x_i]$$

provided that the series converges absolutely. Or in a more abbreviated notation

$$E\mathbf{x} = \sum_X x \cdot p_{\mathbf{x}}(x) = \sum_X x \cdot p(x). \tag{3.1}$$

Absolute convergence of the series $\sum_X xp(x)$ is equivalent to convergence of the positive part $\sum_{x>0} xp(x)$ and the negative part $\sum_{x \le 0} xp(x)$. If $\sum_X |x| p(x)$ diverges, so that $\sum_X x \cdot p(x)$ diverges to $+\infty$ or $-\infty$ or depends on the order of summation, then $E\mathbf{x}$ does not exist.

In our discussion of the game between Peter and Paul we have attempted to make the above definition of expectation seem natural, but we emphasize that the definition could have been otherwise if this mathematical concept had not agreed with the intuitive meaning of the word expectation. The expectation concept seems to be due to Huygens (1657) who asserts in the third proposition of his treatise that if a player has p chances of gaining a and q chances of gaining b, his expectation is $(pa + qb)/(p + q)$.

Much later Daniel Bernoulli maintained that, in one instance at least, the so-called Petersburg paradox, the expectation concept is directly opposed to intuition.

The Petersburg paradox is as follows: A flips a coin; if head appears at the first throw he is to receive a shilling from B, if not he flips the coin again and receives 2 shillings if a head results. In general, A receives 2^{n-1} shillings if a head appears for the first time on the nth flip. The expectation is

$$\sum_{n=1}^{\infty} 2^{n-1} \cdot \frac{1}{2^n}$$

which diverges to $+\infty$ so that A ought to be willing to pay any sum to persuade B to play with him. But this is nonsense as it is very likely that A's return will be small.

The "paradox" has been explained in several ways and Huygens' definition of expectation is now universally accepted. One explanation involves the economic concept of decreasing marginal utility: the worth to A of winning an additional shilling is a decreasing function of his total winnings. Thus, if w and x are worth and winnings and if, for example, $\Delta w = k\,\Delta x/x$, then $w = k \log x$, where x is measured in appropriate units. Now the expected worth to A of playing the game is

$$\sum_{n=1}^{\infty} k \log 2^{n-1} \cdot \frac{1}{2^n} = k \log 2$$

which is more in agreement with intuition.

It often occurs that we will have the probability function of an r.v. $\mathbf{x}(s)$, but we want to know the expectation of some related r.v. $\mathbf{y}(s) = z(\mathbf{x}(s))$; thus, knowing $p_{\mathbf{x}}(x)$, we may desire $E\mathbf{x}^2$. The following result is useful in these cases:

Theorem 3.1

The expectation of the r.v. $\mathbf{y}(s) = z[\mathbf{x}(s)]$ is given by $E\mathbf{y} = \sum_X z(x)p_{\mathbf{x}}(x)$.

Proof

Write Y for the range of $\mathbf{y}$. According to the definition

$$E\mathbf{y} = \sum_Y y \cdot p_{\mathbf{y}}(y).$$

In order that $\mathbf{y}(s)$ should be a bonafide r.v., $z(x)$ must be a single-valued function.

Therefore

$$p_{\mathbf{y}}(y) = P\{z[\mathbf{x}(s)] = y\} = \sum_{\{x:\, z(x) = y\}} P[\mathbf{x}(s) = x]$$

and

$$\sum_Y y p_{\mathbf{y}}(y) = \sum_Y y \sum_{\{x:\, z(x) = y\}} p_{\mathbf{x}}(x) = \sum_Y \sum_{\{x:\, z(x) = y\}} z(x)p_{\mathbf{x}}(x) = \sum_X z(x)p_{\mathbf{x}}(x).$$

Theorem 3.2

If a and b are constants while $\mathbf{x}$ and $\mathbf{y}$ are r.v.'s whose expectations exist then

$$E(a\mathbf{x}) = a \cdot E\mathbf{x} \tag{3.2}$$

$$E(b + \mathbf{x}) = b + E\mathbf{x} \tag{3.3}$$

$$E(\mathbf{x} + \mathbf{y}) = E\mathbf{x} + E\mathbf{y} \tag{3.4}$$

that is, E is a linear operator.

Proof (for discrete r.v.'s)

Denote the ranges of $\mathbf{x}$ and $\mathbf{y}$ by $X = \{x_1, x_2, \ldots\}$ and $Y = \{y_1, y_2, \ldots\}$, respectively.

$$E(a\mathbf{x}) = \sum_X axp(x) = a \sum_X xp(x) = aEx,$$

which is Equation (3.2). Equation (3.3) is proved in a similar manner. Equation (3.4) is demonstrated as follows:

$$\begin{aligned} E(\mathbf{x} + \mathbf{y}) &= \sum_X \sum_Y (x + y)P(\mathbf{x} = x, \mathbf{y} = y) \\ &= \sum_X x \sum_Y P(\mathbf{x} = x, \mathbf{y} = y) + \sum_Y y \sum_X P(\mathbf{x} = x, \mathbf{y} = y) \\ &= \sum_X xp(x) + \sum_Y yp(y) = E\mathbf{x} + E\mathbf{y}. \end{aligned}$$

(The first step in the above chain of equalities uses an easy generalization of Theorem 3.1 to functions of two r.v.'s.)

The necessity of the existence of $E\mathbf{x}$ and $E\mathbf{y}$ in the statement of Theorem 3.2 is shown by the following example. Let

$$p(s_i) = \frac{1}{i} - \frac{1}{i + 1},$$

$\mathbf{x}(s_i) = i$, and $\mathbf{y}(s_i) = 1 - i$ for $i = 1, 2, \ldots$. We have $E(\mathbf{x} + \mathbf{y}) = 1$ but attempting to calculate $E\mathbf{x}$:

$$\sum_{i=1}^{\infty} i\left(\frac{1}{i} - \frac{1}{i + 1}\right) = \frac{1}{2} + \frac{1}{3} + \cdots$$

which diverges to $+\infty$; similarly $E\mathbf{y}$ diverges to $-\infty$. Thus, the left-hand side of Equation (3.4) is 1, while the right-hand side does not exist. On the other hand, there is no reason why the conclusion of Theorem 3.2 should hold since the hypothesis is violated.

Two r.v.'s $\mathbf{x}$ and $\mathbf{y}$ will be called *independent* if $P[(\mathbf{x} \in A) \cap (\mathbf{y} \in B)] = P(\mathbf{x} \in A) \cdot P(\mathbf{y} \in B)$ for any sets A and B in the ranges of $\mathbf{x}$ and $\mathbf{y}$, respectively. A finite number of r.v.'s $\mathbf{x}_1, \ldots, \mathbf{x}_n$ are independent if

$$P[(\mathbf{x}_{i_1} \in A_{i_1}) \cap \cdots \cap (\mathbf{x}_{i_k} \in A_{i_k})] = P(\mathbf{x}_{i1} \in A_{i_1}) \cdots P(\mathbf{x}_{i_k} \in A_{i_k})$$

for every subset $i_1, \ldots, i_k$ of the first n integers.

Not all functions will serve as probability functions; Kolmogorov's axioms impose certain restrictions. We call attention to the following *Criteria.* If $p(x)$ is an arbitrary function defined on a discrete set $X = \{x_1, x_2, \ldots\}$ and if $p(x)$ satisfies

$$p(x) \geq 0, \quad x \in X \tag{3.5}$$

$$\sum_X p(x) = 1, \tag{3.6}$$

then $P(A)$ defined by

$$P(A) = \sum_{x \in A} p(x) \quad \text{for each } A \subset X \tag{3.7}$$

satisfies the axioms (with X acting as the certain event), and hence, is a probability measure.

3.2 THE BINOMIAL PROBABILITY FUNCTION

We now return to the probabilistic experiment of observing whether a flipped thumb tack falls point up or point down. If x denotes the number of times, out of n flips, that the tack falls point up then we may ask for the probability function of x. Or, what amounts to the same thing, we may ask: What is the probability $b(x; n, p)$[3] of exactly x successes in a sequence of n Bernoulli trials, the probability of success on a single trial being p? This is easily solved as follows: The probability of x successes out of n, in any particular order, is $p^x(1-p)^{n-x}$, $0 \leq x \leq n$. Thus, since the number of ways of ordering x S's and $n - x$ F's is $\binom{n}{x}$, then

$$b(x; n, p) = \binom{n}{x} p^x (1-p)^{n-x}; \quad x = 0, 1, \ldots, n. \tag{3.8}$$

[3] Probability functions are designated by the first letter of their names. Thus, we have $b(x; n, p)$ for the binomial and later $h(x; n, D, N)$ and $p(x; \lambda)$ for the hypergeometric and Poisson probability functions.

This is the *binomial probability function.* Using the criteria of (3.5) and (3.6), we may check that $b(x; n, p)$ can in fact serve as a probability function.

If a coin is flipped 100 times then nearly everyone will "expect" 50 heads. Let us see if this result is consistent with our formal definition of expectation. From Equations (3.1) and (3.8), we have as the expected number of successes in n Bernoulli trials:

$$E\mathbf{x} = \sum_{x=0}^{n} x\binom{n}{x}p^x(1-p)^{n-x}.$$

It is instructive to evaluate this expectation directly, and the reader is encouraged to do so; however, we will perform an indirect evaluation which illustrates the use of Equation (3.4). Define the "indicator" variables

$$\mathbf{x}_i = \begin{cases} 1 \text{ if } S \text{ occurs on the } i\text{th trial} \\ 0 \text{ if } F \text{ occurs on the } i\text{th trial.} \end{cases}$$

Now $\mathbf{x} = \mathbf{x}_1 + \cdots + \mathbf{x}_n$ is the total number of successes in n Bernoulli trials. $E\mathbf{x}_i = 1 \cdot p + 0 \cdot (1 - p) = p$ and $E\mathbf{x} = E\mathbf{x}_1 + \cdots + E\mathbf{x}_n$ so that

$$E\mathbf{x} = n \cdot p. \tag{3.9}$$

Taking $n = 100$ and $p = \frac{1}{2}$, we see that in this sense one should indeed expect 50 heads in 100 flips of a coin.

3.3 THE POISSON PROBABILITY FUNCTION

This section introduces the Poisson probability function which is important as an approximation to other probability functions, notably the binomial and the hypergeometric to be introduced in the next section.

The Poisson probability function is also important for its own sake. Let us consider the collisions of a distinguished molecule of gas with the many other molecules making up a gas in steady state. We might ask for the probability $P_t(x)$ of x collisions in time t. Problems of this type are important in the study of heat. Between collisions, the distinguished molecule will travel in a straight line at a constant speed. Hence, it is not unreasonable to assume that in a short interval of time $(t, t + \Delta t)$, the probability of a single collision is proportional to Δt while the probability of two or more collisions vanishes to an order higher than Δt. Further, because the gas is in a steady state, ρ the constant of proportionality will be independent of t and collisions in disjoint time intervals will not influence one another. To be precise, our assumptions are as follows:

(i) Events defined on nonoverlapping time intervals are statistically independent

(ii) $P_{\Delta t}(0) = 1 - \rho \cdot \Delta t + o(\Delta t)$

(iii) $P_{\Delta t}(1) = \rho \cdot \Delta t + o(\Delta t)$

where $o(\Delta t)$ denotes a quantity which approaches 0 faster than Δt. To evaluate $P_t(x)$, we divide the interval $(0, t)$ into n equal parts. Neglecting quantities which approach 0 as $n \to \infty$, we have a sequence of n Bernoulli trials. Hence for $x = 0, 1, 2, \ldots$

$$P_t(x) = \lim_{n \to \infty} \left[\binom{n}{x} \left(\rho \frac{t}{n} \right)^x \left(1 - \rho \frac{t}{n} \right)^{n-x} \right] \tag{3.10}$$
$$= \frac{(\rho t)^x}{x!} e^{-\rho t}.$$

In general, an r.v. $\mathbf{x}$ will be said to have the *Poisson distribution* with parameter λ if

$$P(\mathbf{x} = x) = p(x; \lambda) = \frac{\lambda^x}{x!} e^{-\lambda}, \quad x = 0, 1, 2, \ldots. \tag{3.11}$$

We denote the Poisson probability function (3.11) by $p(x; \lambda)$. It is easy to show, using the series expansion for the exponential function, that

$$\sum_{x=0} p(x; \lambda) = 1,$$

and hence, $p(x; \lambda)$ can represent a probability. A related computation provides us with an interpretation of the parameter λ. We have

$$E\mathbf{x} = \sum_{x=0}^{\infty} x p(x; \lambda)$$
$$= \lambda \sum_{x=1}^{\infty} \frac{\lambda^{x-1}}{(x-1)!} e^{-\lambda}$$
$$= \lambda \sum_{y=0}^{\infty} p(y; \lambda) = \lambda.$$

The expectation of a Poisson r.v. is λ.

Our example, concerning the molecule of gas, shows that if $\lambda = np$ is a constant then the binomial approaches the Poisson:

$$\lim_{n \to \infty} b(x; n, p) = p(x; \lambda).$$

This result suggests, but does not prove, that for n large and p fixed but small $b(x; n, p)$ should be well approximated by $p(x; np)$. In fact, this approximation works quite well.[4]

3.4 APPLICATION—QUALITY CONTROL ATTRIBUTE SAMPLING

Suppose it is desired to determine by inspection whether a lot of N items (say water pumps) all manufactured under the same conditions is of acceptable quality or not. Since the process of inspection may be costly or even destructive, we may wish to inspect some subset of the entire lot. Let us examine the consequences of adopting the following scheme: From the lot, choose a sample of n items; if d or fewer are defective then accept the lot of N, otherwise reject the lot as being of unacceptable quality. d is called the allowable number of defectives. This inspection scheme is characterized by the three quantities (N, n, d). Let the random variable $\mathbf{x}$ denote the number of defectives in the sample. If there are D defectives in the lot and the sampling has been done at random then the probability that $\mathbf{x} = x$ is

$$h(x; n, D, N) = \frac{\binom{D}{x}\binom{N - D}{n - x}}{\binom{N}{n}}; \quad x = 0, 1, \ldots, n. \tag{3.12}$$

This is the *hypergeometric* probability function which often arises in practice but is infrequently used since the needed tables are voluminous and, as we now show, adequate approximations are available.

Theorem 3.3

If $N \to \infty$ while $D/N \to p$, a constant, then

$$h(x; n, D, N) \sim b(x/n, p)$$

in the sense that the ratio of the two sides approaches 1.

This theorem is reasonable since, when N is large relative to n, the proportion defective for the second and later items sampled is changed but little by the choice of the first item.

[4] Feller, William, *An Introduction to Probability Theory and Its Applications I*, Second Edition, Wiley, New York, 1957, p. 161.

Proof

From the relations $D \sim Np$ and $N - D \sim N(1 - p)$ we see that D and $N - D$ approach ∞ with N. Next, writing $a^{(b)} = a(a-1)(a-2)\cdots(a-b+1)$, we see that

$$h(x; n, D, N) = \binom{n}{x} \frac{D^{(x)}(N - D)^{(n-x)}}{N^{(n)}}.$$

But for fixed b, $\lim_{a \to \infty} a^{(b)}/a^b = 1$, so that $h(x; n, D, N) \sim \binom{n}{x} p^x(1 - p)^{n-x}$.

It is customary in the quality control application to employ the additional Poisson approximation of the binomial. These two approximations are particularly appropriate for the present application since they require the sample and lot sizes to be large while the proportion of defectives is small. Thus, frequently we may approximate $h(x; n, D, N)$ by $p(x; nD/N)$.

Returning to the quality control sampling scheme (N, n, d) and using the Poisson approximation, we see that the probability of rejecting the lot is

$$R(p) = \sum_{x=d+1}^{\infty} p(x; \lambda)$$

where $p = D/N$ and $\lambda = np$. The consequences of adopting any particular scheme may be shown pictorially by a graph of R, the probability of rejection, versus p, the proportion of defectives in the lot. The function $R(p)$ is called the Operating Characteristic (O.C.) curve. A typical O.C. curve is shown in Figure 3.1.

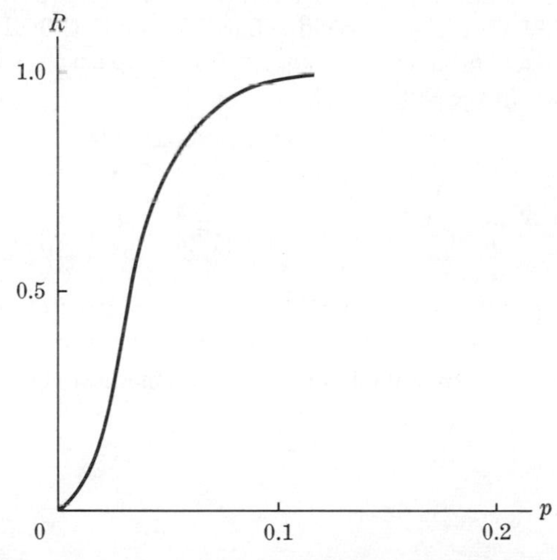

Figure 3.1 O.C. Curve for the Scheme (1000, 100, 3)

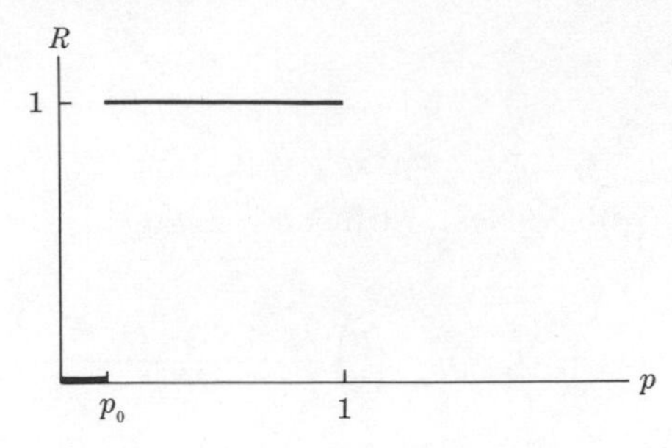

Figure 3.2 Ideal O.C. Curve

If we are willing to tolerate p_0 (or less) proportion defective then ideally we would like the O.C. curve to look as it does in Figure 3.2. The ideal curve can be achieved only through complete inspection which may be uneconomical in many cases. We may decide between several possible sampling schemes by comparing their O.C. curves with the ideal.

*3.5 APPLICATION—A PROBLEM IN MAXIMUM LIKELIHOOD ESTIMATION

The starting point for this section is a problem of Moroney.[5]

The amount of dust in the atmosphere may be estimated by using an ultramicroscope. A very small volume of air is illuminated by a spark and the observer counts the number of particles of dust he sees. By repeating this operation a large number of times, the amount of dust in each cubic centimeter of air can be estimated. Suppose that the following test results were obtained in a series of 300 spot checks by the flash method. Calculate the expected frequencies for each number of particles for comparison with the observed frequencies shown in the table.

Number of particles	0	1	2	3	4	5	More than 5
Frequency of occurrence	38	75	89	54	20	19	5

Moroney has in mind that the 300 tests are independent probability experiments each obeying the same Poisson probability law. Thus if $\mathbf{x}_i$ is the number of particles observed on the ith flash then the joint probability function of $\mathbf{x}_1, \mathbf{x}_2, \ldots, \mathbf{x}_{300}$ is

$$P(\mathbf{x}_1 = x_1, \mathbf{x}_2 = x_2, \ldots, \mathbf{x}_{300} = x_{300}) = \frac{\lambda^{\sum_1^{300} x_i}}{x_1!\, x_2! \cdots x_{300}!} e^{-300\lambda}. \quad (3.13)$$

[5] Moroney, M. J., *Facts from Figures*, Second Edition, Penguin Books, London, 1954.

This function gives the probabilities of various test results if one knows the probability function, but here we are concerned with a reverse question. What probability function is likely to have yielded the observed test results? Because of the Poisson probability assumption this reduces to estimating the parameter λ. R. A. Fisher has suggested using the probability function (3.13) in the following reverse manner. Considering $x_1, x_2, \ldots, x_{300}$ to be the observed results of a series of tests, write

$$L(\lambda) = \frac{\lambda^{\sum_{1}^{300} x_i}}{x_1!\, x_2! \cdots x_{300}!} e^{-300\lambda}.$$

Now estimate λ by maximizing $L(\lambda)$. The result is

$$\hat{\lambda} = \sum_{1}^{300} \frac{x_i}{300}.$$

This would solve Moroney's problem if the frequencies of occurrence were not grouped for more than five particles. Considering "more than 5" to be 6, we find $\hat{\lambda} = 620/300 \simeq 2.07$. No doubt this is a perfectly good common sense solution, but we are dissatisfied with the method.

To determine the likelihood of the so-called truncated sample we must determine $A = P\{n \text{ out of } N \text{ observations are } < K \text{ and equal to } x_1, x_2, \ldots, x_n.\}$ For Moroney's problem $n = 295$, $N = 300$, and $K = 6$. Now, adopting an abbreviated terminology,

$$\begin{aligned} A &= P\{n \text{ out of } N < K\} \cdot P\{x_1, x_2, \ldots, x_n | n \text{ out of } N < K\} \\ &= \binom{N}{n} P^n (1 - P)^{N-M} \prod_{i=1}^{n} \left(\frac{\lambda^{x_i} e^{-\lambda}}{x_i!\, P} \right) \end{aligned}$$

where $P = P(\mathbf{x}_i < K) = \sum_{i=0}^{K-1} \lambda^i e^{-\lambda}/i!$.

On differentiating log A with respect to λ and equating to 0, we obtain

$$\frac{-P'}{1 - P}(N - n) + \frac{\sum_{i=1}^{n} x_i}{\lambda} = n.$$

Substituting

$$\begin{aligned} \frac{-P'}{1 - P} &= \frac{K}{\lambda}\left(1 + \frac{\lambda}{K + 1} + \frac{\lambda^2}{(K + 1)(K + 2)} + \cdots\right)^{-1} \\ &= \frac{K}{\lambda}\left(1 - \frac{\lambda}{K + 1} + \frac{\lambda^2}{(K + 1)^2(K + 2)} - \cdots\right) \end{aligned}$$

yields

$$\frac{n(1+K)}{n+NK}\left(\frac{\sum_{i=1}^{n} x_i}{n}+\frac{N-n}{n}K\right) \le \hat{\lambda} \le \frac{\sum_{i=1}^{n} x_i}{n}+\frac{N-n}{n}K,$$

with the lower bound being the more accurate second approximation. Moroney's data yields (0.986) (2.10) $\le \hat{\lambda} \le$ 2.10 and $\hat{\lambda} \simeq$ 2.07 for the truncated sample.

PROBLEMS

1. (de Méré's problem). A game between two persons is won by the player who first scores three points. The participants place equal stakes and the winner takes all. If the players decide to quit when they have one and two points respectively and their chances for winning each point are equal, what should be the division of stakes?

2. $\mathbf{x}$ is an r.v. with probability function $p(x)$ tabulated below.

x	-2	-1	0	1	2
$p(x)$	$\frac{1}{8}$	$\frac{3}{8}$	0	$\frac{3}{8}$	$\frac{1}{8}$

Compute the mean of $\mathbf{x}$. Is it a possible value of $\mathbf{x}$?

3. A coin is drawn at random from a bag containing three pennies, a nickel and two dimes. What is the expected amount of the draw?

4. Using the continuity property of probability, prove that if $P(|\mathbf{x} - \mu| < \epsilon) = 1$ for every $\epsilon > 0$ then $P(\mathbf{x} = \mu) = 1$.

5. Let $\mathbf{x}_1$ and $\mathbf{x}_2$ be independent r.v.'s assuming values on the positive integers. Under what conditions is the event $\{\mathbf{x}_1 \text{ is even}\}$ independent of the event $\{\mathbf{x}_1 + \mathbf{x}_2 \text{ is even}\}$?

6. An r.v. $\mathbf{x}$ has probability function $P[\mathbf{x} = (-1)^{k-1}2^k/k] = 2^{-k}$, $k = 1, 2, \ldots$. What is $E\mathbf{x}$?

7. An age breakdown of the 100 lions in a certain zoo is given below. If these *proportions* remain constant as time passes then what is the life expectancy of the resident lions?

Age	0	1	2	3	4	5
Number of Lions	30	28	22	12	6	2

What is the remaining life expectancy of a lion of age 2?

8. Using the mortality Table 1.2, determine the age at which a newborn child may expect to die. Determine a fair price for a three-year life insurance policy in the amount of $1,000 for a man who is 50 years old. (The estate's expected gain should be 0.)

9. In Problem 1.12, what is the expected value of the drawn coin?

10. One boy has six coins, another five. They match coins until one boy has won them all. Show that they may expect 30 tosses.

11. A game of ping-pong is won by the player who first wins 21 points and is ahead by two. If a player has probability 0.49 of winning a point, what is the probability that he will upset his opponent by winning the game?

12. A game of tennis is won by the player who first wins four points and is ahead by two. A set is won by the player who first obtains six games and is ahead by two games. A match is the best out of three sets. If a player's probability of winning a point is 0.49 then what are his chances of upsetting his opponent in the match?

13. (Bernstein) Given three random variables x, y, and z, in (x, y, z) space, each of the points (1, 0, 0), (0, 1, 0), (0, 0, 1), and (1, 1, 1) carry probability $\frac{1}{4}$. Show

$$P(\mathbf{x} = 1) = P(\mathbf{y} = 1) = P(\mathbf{z} = 1) = \tfrac{1}{2}$$

and

$$\begin{aligned} P(\mathbf{x} = 1, \mathbf{y} = 1) = P(\mathbf{x} = 1, \mathbf{z} = 1) &= P(\mathbf{y} = 1, \mathbf{z} = 1) \\ &= P(\mathbf{x} = 1, \mathbf{y} = 1, \mathbf{z} = 1) = \tfrac{1}{4}. \end{aligned}$$

What are the implications for independence?

14. (Crow) Let the points (1, 0, 0), (0, 1, 0), (0, 0, 1), (1, 1, 0), and (1, 1, 1) carry probabilities $\frac{1}{8}, \frac{1}{8}, \frac{3}{8}, \frac{2}{8}$, and $\frac{1}{8}$, respectively. Discuss the significance of

$$P(\mathbf{x} = 1, \mathbf{y} = 1, \mathbf{z} = 1) = P(\mathbf{x} = 1)\cdot P(\mathbf{y} = 1)\cdot P(\mathbf{z} = 1)$$

but

$$P(\mathbf{x} = 1, \mathbf{y} = 1) \neq P(\mathbf{x} = 1)\cdot P(\mathbf{y} = 1).$$

15. You play a game with an opponent. The game proceeds as follows: You and your opponent toss a coin (the same one) alternately in turn. The first person to toss heads wins. You start the game. Assume the tosses are independent and that the probability of tossing heads on any one trial is p.

    (a) Find the probability that you win the game.

    (b) Suppose you gain $1 if you win and gain $-$$1 if you lose. What is your expected gain if you play this game once?

16. A man claims he can pick the winning horse in a race 90% of the time. To test his claim he picks horses to win in 10 races. Assume the results to be Bernoulli trials.

    (a) What is the probability of his picking at least 9 out of 10 winners if indeed his probability of picking a winner on any one race is 0.9?
    (b) Suppose the man is guessing and there are 10 horses in each race for which he chooses at random. What is the probability he will pick at least 5 races correctly?
    (c) If he guesses, which is the most probable number of correct choices?
    (d) If he guesses, what is his expected number of correct choices?

17. 100 people are subject to a blood test. The blood samples of 10 people will be pooled and analyzed together. If the test is negative, one test suffices for the 10 people. If the test is positive, each of the 10 persons must be tested separately and, in all, 11 tests are required for the 10 people. Assume the probability that the test is positive is 0.1 for all people and that the people are stochastically independent. Let $\mathbf{x}$ be the number of tests necessary for the 100 people. Find $E\mathbf{x}$.

18. Let $\mathbf{x}$ have the *geometric* or *Pascal* probability function

    $$P(\mathbf{x} = x) = g(x; q) = (1 - q)q^x, \quad x = 0, 1, 2, \ldots, \text{ where } 0 < q < 1.$$

    Show that the conditional probability satisfies the equation

    $$P(\mathbf{x} = k + x | x \geq k) = P(\mathbf{x} = x), \quad x = 0, 1, 2, \ldots, \quad k = 0, 1, 2, \ldots.$$

19. Let $\mathbf{x}$ be an r.v. whose range is the set of all nonnegative integers and which satisfies

    (a) $0 < P(\mathbf{x} = 0) < 1$, and
    (b) $P(\mathbf{x} = k + x | \mathbf{x} \geq k) = P(\mathbf{x} = x)$ for all nonnegative integers x, k.

    Then prove that $\mathbf{x}$ has the Pascal distribution where $P(\mathbf{x} = 0) = 1 - q$.

20. Let $\mathbf{x}_1, \ldots, \mathbf{x}_n$ be n independent identically distributed r.v.'s each having the geometric probability function. Prove that $\mathbf{y} = \min(\mathbf{x}_1, \ldots, \mathbf{x}_n)$ has probability function $P(\mathbf{y} = y) = g(y; q^n)$.

21. Let the probability that a family has exactly n children be μp^n for $n \geq 1$. Suppose, on all births, that the sexes are equally likely and independent. Show the probability that a family contains k boys is

    $$\frac{2\mu p^k}{(2 - p)^{k+1}}, \quad k \geq 1.$$

22. A newsboy purchases papers at four cents and sells them at seven but he may not return unsold papers. Daily demand for papers follows a Poisson distribution with mean 50. How many papers should he purchase in order to maximize his expected profit?

23. An air force is considering an attack on a target of value Ta where a is the value of an airplane. No matter what size the flight, each plane has a probability p_1 of being shot down before reaching the target and planes reaching the target each have a probability p_2 of being shot down before reaching their base. If k planes reach the target then the probability of destroying it is $k/(k + 1)$. Find the expected net gain from sending n planes in the attack. Assuming $p_1 = p_2 = p$, what is the best strategy if p is small enough that p^2 can be neglected?

4

PROBABILITY DISTRIBUTION FUNCTIONS AND CONTINUOUS RANDOM VARIABLES

4.1 DISTRIBUTION FUNCTIONS

In our discussion of the nature of probability we introduced the concept of a probabilistic experiment and discussed several elementary examples. A more complicated probabilistic experiment is provided by setting an alarm clock and measuring the alarm timing error which equals the actual time when the bell first rings minus the time for which the alarm is set. If the measuring instruments are accurate to the nearest tenth of a second, and if we think of the alarm error, in seconds, as being plotted on a line, then the simple events will be all multiples of $\frac{1}{10}$ within a certain interval, say $+300$ to -300. Since the length of this interval is long compared to the accuracy of the instruments (600 versus $\frac{1}{10}$), it is convenient to idealize this experiment to include *all* points on the line as simple events. This kind of idealization is essential in many situations to obtain a model which is of manageable proportions. The model is particularly appropriate since if the measuring accuracy is improved to say a hundredth of a second then all multiples of $\frac{1}{100}$ between $+300$ and -300 actually are possible outcomes of the experiment. In passing, note that if the model is to be an accurate approximation to the true situation, it will be necessary to assign a very small amount of probability to that portion of the line which is outside the interval $+300$ to -300.

For "error of measurement" problems of the above type it is frequently possible to describe the random variation of the error magnitude $\mathbf{x}$ in the following manner:

$$P(\mathbf{x} \leq x) = \int_{-\infty}^{x} \frac{1}{\sqrt{2\pi}\sigma} \exp\left[-\frac{(t-\mu)^2}{2\sigma^2}\right] dt.$$

This is, of course, the famous normal or Gaussian probability integral. At a later point, we will see that the apparent complexity of this expression is mostly illusory. For now it will suffice to say that μ and σ are parameters having to do with location and scale, respectively. Various probabilities of interest can be obtained from $P(\mathbf{x} \leq x)$; we have, for instance,

$$P(\mathbf{x} > x) = 1 - P(\mathbf{x} \leq x)$$

and

$$P(|\mathbf{x}| \leq a) = P(\mathbf{x} \leq a) - P(\mathbf{x} < -a).$$

The purpose of the preceding discussion has been to motivate the following general definitions. A *real-valued random variable* is a function whose domain is an event space S and whose range is the real line. Discrete r.v.'s, treated in Chapter 3, are a special kind of real-valued r.v. Henceforth, we will understand that all r.v.'s are real valued unless it is stated otherwise. The function $F(x) = P(\mathbf{x} \leq x)$ will be called the *distribution function* (d.f.) of the random variable $\mathbf{x}$. A few properties of distribution functions are almost immediately apparent. First, from the nonnegativity of the probability function we have for $y > x$

$$F(y) = F(x) + P(x < \mathbf{x} \leq y) \geq F(x) \tag{4.1}$$

so that all d.f.'s are monotone nondecreasing.

Next, let $y_1 > y_2 > \cdots$ be a sequence approaching x from the right and write $B_i = \{s: \mathbf{x} \leq y_i\}$ then $B_1 \supset B_2 \supset \cdots$ and $\bigcap_{i=1} B_i = \{s: \mathbf{x} \leq x\}$. From the continuity property of probability we see that

$$\lim_{y_i \to x_+} F(y_i) = F(x). \tag{4.2}$$

Thus, every d.f. is continuous from the right. Examples of d.f.'s which are not everywhere continuous from the left are common; for example, the d.f. of every discrete r.v. will have points of discontinuity. Arguments similar to those used to establish Equation (4.2) yield

$$\lim_{x \to \infty} F(x) = 1 \quad \text{and} \quad \lim_{x \to -\infty} F(x) = 0. \tag{4.3}$$

The reader is encouraged to draw graphs of the following simple d.f.'s:

$$P(\mathbf{x} \leq x) = \begin{cases} 0, & x < 0 \\ \frac{1}{2}, & 0 \leq x < 1 \\ 1, & 1 \leq x \end{cases} \tag{4.4}$$

$$P(\mathbf{x} \leq x) = \begin{cases} 0, & x < 0 \\ x, & 0 \leq x < 1 \\ 1, & 1 \leq x \end{cases} \tag{4.5}$$

$$P(\mathbf{x} \le x) = \frac{1}{\sqrt{2\pi}} \int_{-\infty}^{x} e^{-t^2/2}\, dt, \quad -\infty < x < \infty. \tag{4.6}$$

From the definition of a d.f., it is clear that for d.f. (4.4)

$$P(\mathbf{x} = x) = \begin{cases} \frac{1}{2} & x = 0 \text{ or } 1 \\ 0 & \text{otherwise} \end{cases}$$

while for d.f.'s (4.5) and (4.6), $P(\mathbf{x} = x) = 0$, $-\infty < x < \infty$. Positive probability will be assigned to a point, x_0, if and only if the d.f. is discontinuous at x_0. The probability assigned to x_0, which we have called the probability function $p(x_0)$, is the magnitude of the "jump" at this point. Probability functions have been found to be very useful in connection with discrete r.v.'s; $p(x)$ would in fact be sufficient for discussing (4.4). However, for (4.5) and (4.6) the probability function (height of jump) is everywhere zero, and thus would not be a satisfactory tool for discussing probability. What is needed here is the probability density function.

4.2 DENSITY FUNCTIONS

The distribution function $F(x)$ of the r.v. $\mathbf{x}$ will be called *smooth*, if (i) $F(x)$ is continuous on the entire real line, and (ii) $F(x)$ has a continuous derivative except at a finite number of points. For smooth d.f.'s the *density function* of $\mathbf{x}$ is defined to be the derivative of $F(x)$ where it exists:

$$f(x) = \frac{dF(x)}{dx} = F'(x). \tag{4.7}$$

If

$$F'(x) = \lim_{\Delta x \to 0} \frac{F(x + \Delta x) - F(x)}{\Delta x}$$

is infinite, or not unique at a point, then the density is not determined there. The distributions (4.4), (4.5), and (4.6) yield easy examples: (4.4) is discrete, and hence, not smooth; (4.6) is smooth with density

$$(2\pi)^{-1/2} \exp\left(-\frac{x^2}{2}\right), \quad -\infty < x < \infty;$$

(4.5) is smooth with density function

$$f(x) = \begin{cases} 0 & x < 0 \quad \text{or} \quad x > 1 \\ 1 & 0 < x < 1 \\ \text{undetermined} & x = 0, 1. \end{cases}$$

The mean value theorem provides an interpretation of the density function. In an interval (a, b) within which $F(x)$ has a continuous derivative

$$P(a < \mathbf{x} \leq b) = F(b) - F(a) = f(\xi)(b - a) \tag{4.8}$$

where $a < \xi < b$. In this sense, $P(x < \mathbf{x} \leq x + \Delta x) \simeq f(x) \cdot \Delta x$.

The density is the derivative of the distribution, so the distribution will be the integral of the density function. In more detail, partition the interval (c, x) as shown in Figure 4.1.

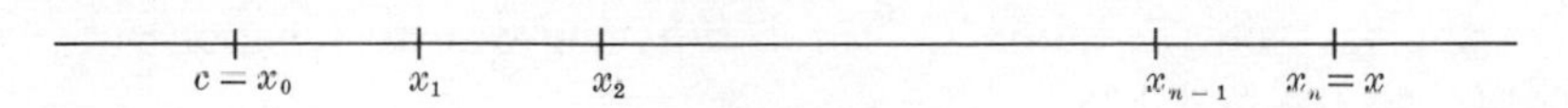

Figure 4.1

Then

$$F(x) = \sum_{i=1}^{n} [F(x_i) - F(x_{i-1})] + F(c).$$

Taking care to include points at which $F(x)$ is not continuously differentiable among the partition points, we have from (4.8)

$$F(x) = \sum_{i=1}^{n} f(\xi_i)(x_i - x_{i-1}) + F(c)$$

where $x_{i-1} < \xi_i < x_i$. Since these are the sums used in defining the Riemann integral of ordinary calculus,

$$F(x) = \int_c^x f(y)\, dy + F(c).$$

Now, allowing c to approach $-\infty$, (4.3) yields

$$F(x) = \int_{-\infty}^{x} f(y)\, dy. \tag{4.9}$$

Taken together, the discrete and the smooth distributions are adequate for almost all probability applications but there are other possibilities. For example, if $D(x)$ is discrete while $S(x)$ is smooth, then

$$F(x) = \tfrac{1}{2}D(x) + \tfrac{1}{2}S(x)$$

is neither discrete nor smooth.

The smooth distributions are an important special case of those possessing density functions. If the d.f. of $\mathbf{x}$ is representable in the form

$$F(x) = P(\mathbf{x} \leq x) = \int_{-\infty}^{x} f(y)\, dy$$

where $f \geq 0$, then $\mathbf{x}$ is called a *continuous r.v.* and $F(x)$ is an *absolutely continuous d.f.* with *density function* $f(x)$. All densities have the properties:

(i) $f(x) \geq 0$.

(ii) $\int_{-\infty}^{\infty} f(x)\, dx = 1$.

(iii) $P(a \leq \mathbf{x} \leq b) = \int_a^b f(x)\, dx$.

In analogy with Equation (3.1), for $\mathbf{x}$ having density function $f(x)$, we define the expectation as

$$E\mathbf{x} = \int_{-\infty}^{\infty} xf(x)\, dx = \lim_{\ell \to -\infty} \int_{\ell}^{0} xf(x)\, dx + \lim_{u \to \infty} \int_0^u xf(x)\, dx \tag{4.10}$$

provided that the two limits on the right are finite. The existence or finiteness of these two limits is equivalent to the absolute convergence of $\int_{-\infty}^{\infty} xf(x)\, dx$. The expectation does not exist if $\int_{-\infty}^{\infty} |x| f(x)\, dx$ diverges.

4.3 THE UNIFORM DISTRIBUTION

In performing a numerical calculation one might encounter a number such as 31.6. Typically this number might be the measure in inches of a length, ℓ. If the usual conventions for rounding have been observed then we will be confident that $31.55 \leq \ell \leq 31.65$ but the exact value of ℓ within this interval will be uncertain. We might be willing to consider ℓ as an r.v. $\boldsymbol{\ell}$ distributed on the interval [31.55, 31.65] with the probability of any subinterval depending only on its length. This assumption is enough to determine the distribution of $\boldsymbol{\ell}$ uniquely. We have the following theorem.

Theorem 4.1

If $\mathbf{x}$ is an r.v. defined on the interval [0, 1] and if $P(x < \mathbf{x} \leq y)$ depends only on the length $y - x$ for all $0 \leq x \leq y \leq 1$, then $\mathbf{x}$ has the distribution

$$P(\mathbf{x} \leq x) = F(x) = \begin{cases} 0, & x \leq 0 \\ x, & 0 \leq x \leq 1 \\ 1, & 1 \leq x. \end{cases}$$

To prove the theorem we need the following lemma.

Lemma

If $z(x)$ is an additive function, that is, $z(x + y) = z(x) + z(y)$, and if it is continuous from the right, then $z(x) = c \cdot x$.

Proof of Lemma

$z(x) = z(x) + z(0)$; $z(0) = 0$. $0 = z(x - x) = z(x) + z(-x)$, $z(-x) = -z(x)$;

therefore, the lemma will be proved when it is demonstrated for positive arguments only. $z(mx) = m \cdot z(x)$ for positive integral m. Also, substituting

$$x = \frac{n}{m}, \quad z(n) = z\left[m\left(\frac{n}{m}\right)\right] = mz\left(\frac{n}{m}\right)$$

so that

$$z\left(\frac{n}{m}\right) = \frac{1}{m} z(n) = \frac{n}{m} z(1)$$

for positive integral n and m. Letting $z(1) = c$, the lemma is proved for rational x. To conclude the proof we must treat the case of irrational positive x. Let $r_1, r_2, \ldots,$ be a decreasing sequence of rationals approaching x. Then, from the right continuity,

$$z(x) = \lim_{i \to \infty} z(r_i) = \lim_{i \to \infty} cr_i = c \cdot x.$$

Proof of Theorem 4.1

For arguments on the unit interval let $P(x < \mathbf{x} \leq y) = z(y - x)$, then

$$z(x + y) = P(0 < \mathbf{x} \leq x + y) = P(0 < \mathbf{x} \leq x) + P(x < \mathbf{x} \leq x + y) = z(x) + z(y).$$

The lemma then implies $z(x) = cx$. Now $F(x) = F(0) + P(0 < \mathbf{x} \leq x) = z(x) = cx$, for $0 \leq x \leq 1$. But the boundary condition $F(1) = 1$ implies $c = 1$, so that $F(x) = x$, for $0 \leq x \leq 1$.

The general *uniform density function* is

$$u(x; a, b) = \begin{cases} (b - a)^{-1}, & a < x < b \\ 0, & \text{otherwise.} \end{cases} \tag{4.11}$$

When we say that an r.v. is distributed at random on the interval (a, b) we mean that it has density $u(x; a, b)$.

4.4 TWO-DIMENSIONAL DISTRIBUTION FUNCTIONS

Let $\mathbf{x}(s)$ and $\mathbf{y}(s)$ be two random variables each having the domain S. The joint *distribution function* of $\mathbf{x}$ and $\mathbf{y}$ is defined to be

$$F(x, y) = P[\mathbf{x}(s) \leq x, \mathbf{y}(s) \leq y].$$

To examine the relationship between joint distributions and the previously defined one-dimensional d.f.'s of **x** and **y**, we need to point out that there is an important correspondence between probability and distribution functions.

Consider the two-dimensional case. It is clear from the definition that F is determined by a knowledge of P. On the other hand it is true, although certainly nontrivial, that a knowledge of F determines P uniquely. The formula

$$P(x_0 < \mathbf{x} \leq x_1, y_0 < \mathbf{y} \leq y_1) = F(x_1, y_1) - F(x_1, y_0) - F(x_0, y_1) + F(x_0, y_0)$$

shows that the probability of any rectangle is determined by $F(x, y)$. It is then plausible that sets which are limits of unions of disjoint rectangles, as suggested by Figure 4.2, have probabilities which are also determined by F.

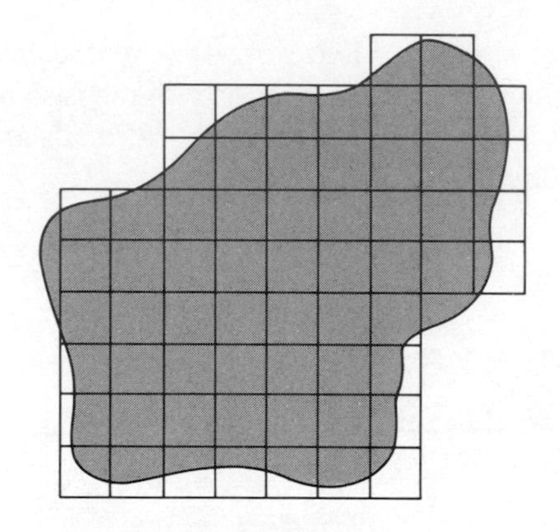

Figure 4.2

Because of the correspondence between distribution and probability functions we may solve problems using whichever formulation is more convenient; if need be, the answer can be converted to the other form.

The joint d.f. determines the previously defined one-dimensional d.f.'s. In fact,

$$\begin{aligned} F_{\mathbf{x}}(x) &= P(\mathbf{x} \leq x) = P(\mathbf{x} \leq x, \mathbf{y} \leq y) + P(\mathbf{x} \leq x, \mathbf{y} > y) \\ &= F(x, y) + P(\mathbf{x} \leq x, \mathbf{y} > y) \end{aligned}$$

for all y. Taking limits as $y \to \infty$ and observing that, according to the continuity property,

$$\lim_{y \to \infty} P(\mathbf{x} \leq x, \mathbf{y} > y) = 0$$

we have

$$F_{\mathbf{x}}(x) = \lim_{y \to \infty} F(x, y), \tag{4.12}$$

and similarly

$$F_{\mathbf{y}}(y) = \lim_{x\to\infty} F(x, y).$$

$F_{\mathbf{x}}(x)$ is called the marginal d.f. of $\mathbf{x}$. The word marginal simply emphasizes that other r.v.'s are present; this is the same distribution of $\mathbf{x}$ that was previously discussed.

If $F(x, y)$ is representable in the form

$$F(x, y) = \int_{-\infty}^{x}\int_{-\infty}^{y} f(u, v)\, dv\, du$$

then $f(x, y)$ is called the *joint probability density function* of $\mathbf{x}$ and $\mathbf{y}$. The density of $\mathbf{x}$ will be

$$f_{\mathbf{x}}(x) = F_{\mathbf{x}}'(x) = \frac{d}{dx}\int_{-\infty}^{x}\int_{-\infty}^{\infty} f(u, v)\, dv\, du$$

$$= \int_{-\infty}^{\infty} f(x, v)\, dv \tag{4.13}$$

and similarly the density of y is given by

$$f_{\mathbf{y}}(y) = \int_{-\infty}^{\infty} f(u, y)\, du.$$

As an example, given

$$f(x, y) = \begin{cases} 1, & 0 < x, y \leq 1 \\ 0, & \text{otherwise} \end{cases}$$

required to determine the joint d.f. of $\mathbf{x}$ and $\mathbf{y}$ and also their marginal distributions.

$$F(x, y) = \begin{cases} \int_0^1\int_0^1 1\, dv\, du = 1; & x \geq 1, y \geq 1 \\ \int_0^1\int_0^y 1\, dv\, du = y; & x \geq 1, 0 \leq y \leq 1 \\ \int_0^x\int_0^1 1\, dv\, du = x; & 0 \leq x \leq 1, y \geq 1 \\ \int_0^x\int_0^y 1\, dv\, du = xy; & 0 \leq x \leq 1, 0 \leq y \leq 1 \\ 0; & x \leq 0 \quad \text{or} \quad y \leq 0 \end{cases}$$

$$F_{\mathbf{x}}(x) = \lim_{y\to\infty} F(x, y) = \begin{cases} 1, & x \geq 1 \\ x, & 0 \leq x \leq 1 \\ 0, & x \leq 0. \end{cases}$$

The r.v.'s $\mathbf{x}$ and $\mathbf{y}$ are each uniformly distributed on the interval [0, 1], and must have the uniform density over this range. We see that the equation

$$F(x, y) = F_{\mathbf{x}}(x) \cdot F_{\mathbf{y}}(y)$$

is satisfied for all x and y. In general, whenever we may calculate the joint d.f. as the product of the marginals, then we say that the r.v.'s $\mathbf{x}$ and $\mathbf{y}$ are independent. This is consistent with previous concepts of independence.

4.5 TRANSFORMATION OF VARIABLES

Often problems of the following type arise in applied probability theory: given the density $f(x, y)$ of the r.v.'s $\mathbf{x}$ and $\mathbf{y}$, required to find the density $g(u, v)$ of the transformed variables $\mathbf{u} = u(\mathbf{x}, \mathbf{y})$ and $\mathbf{v} = v(\mathbf{x}, \mathbf{y})$. If the transformation $u = u(x, y)$, $v = v(x, y)$ is single valued with a single-valued inverse $x = x(u, v)$, $y = y(u, v)$ then under further regularity conditions the relation

$$g(u, v) = f[x(u, v), y(u, v)] \cdot \frac{\partial(x, y)}{\partial(u, v)}. \tag{4.14}$$

is valid where $\partial(x, y)/\partial(u, v)$, called the Jacobian, is the absolute value of the determinant

$$\begin{vmatrix} \dfrac{\partial x}{\partial u} & \dfrac{\partial y}{\partial u} \\ \dfrac{\partial x}{\partial v} & \dfrac{\partial y}{\partial v} \end{vmatrix}.$$

A rigorous derivation of Equation (4.14) becomes complicated, but the basic idea is quite simple. Here we present the idea, but refer the reader to his advanced calculus text[1] for precise conditions and details. Equation (4.14) is in fact a result concerning the transformation of multiple integrals; it states that

$$\iint_R f(x, y)\, dx\, dy = \iint_T f[x(u, v), y(u, v)] \frac{\partial(x, y)}{\partial(u, v)}\, du\, dv \tag{4.15}$$

where $T = \{(u, v): [x(u, v), y(u, v)] \in R\}$.

One proof involves covering the region R, in the (x, y) plane, with a grid indicating various u and v levels, as shown in Figure 4.3. Requiring the grid

[1] For example, Courant, R. *Differential and Integral Calculus II.* Interscience Publishers, Inc., New York, 1949, p. 253.

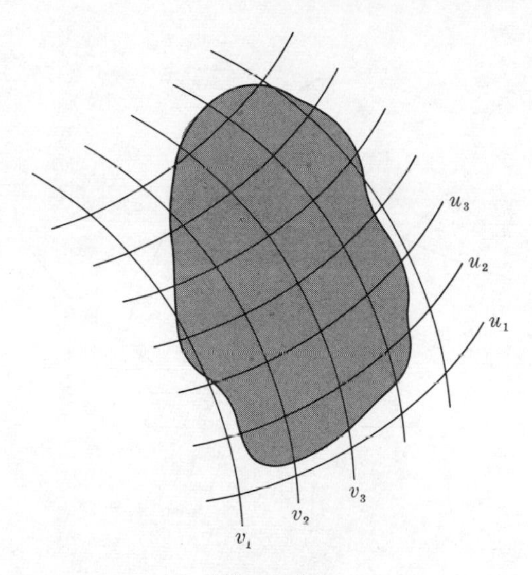

Figure 4.3

to become finer, it is then clear that

$$\iint_R f(x, y)\, dx\, dy = \iint_T f[x(u, v), y(u, v)]\, dA \tag{4.16}$$

where dA is the increment of area of a grid cell.

The differentials of x and y are

$$dx = \frac{\partial x}{\partial u} du + \frac{\partial x}{\partial v} dv$$

$$dy = \frac{\partial y}{\partial u} du + \frac{\partial y}{\partial v} dv$$

hence, approximately for small increments

$$x - x_0 \sim \frac{\partial x}{\partial u} \Delta u + \frac{\partial x}{\partial v} \Delta v$$

$$y - y_0 \sim \frac{\partial y}{\partial u} \Delta u + \frac{\partial y}{\partial v} \Delta v.$$

Therefore, a typical grid cell would have coordinates as indicated in Figure 4.4. But the grid cell is approximately a parallelogram, and we know that the area of a parallelogram with three corners (0, 0), (x_1, y_1), and (x_2, y_2) is the absolute value of

$$\begin{vmatrix} x_1 & y_1 \\ x_2 & y_2 \end{vmatrix}.$$

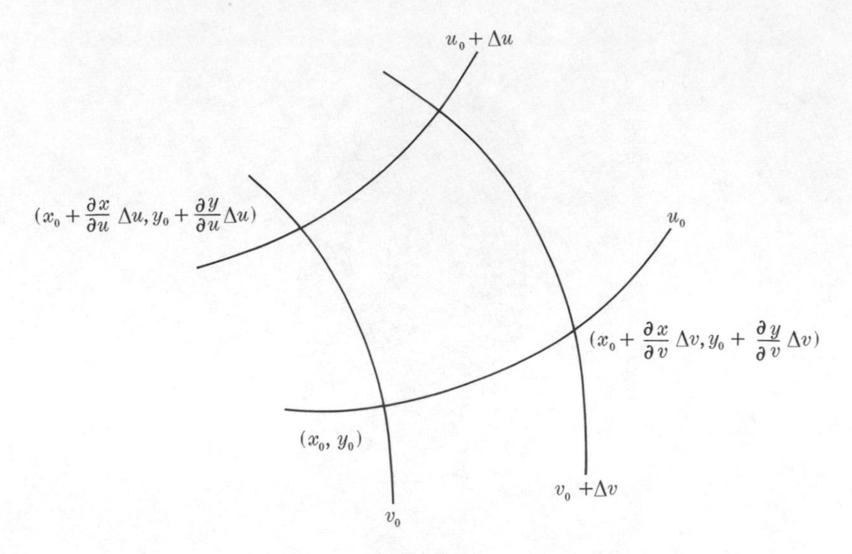

Figure 4.4

Therefore

$$\Delta A = \begin{vmatrix} \frac{\partial x}{\partial u}\Delta u & \frac{\partial y}{\partial u}\Delta u \\ \frac{\partial x}{\partial v}\Delta v & \frac{\partial y}{\partial v}\Delta v \end{vmatrix} = \frac{\partial(x, y)}{\partial(u, v)}\Delta u \, \Delta v$$

and (4.15) follows from (4.16).

Purely as a memory device, one can recall the form of the Jacobian by observing that ∂u and ∂v in the denominator seem to cancel with du and dv in (4.15).

4.6 THE NORMAL DISTRIBUTION

We have already stated, in our introductory discussion, that errors of measurement frequently follow a normal distribution. That is, the error magnitude **x** will have density function

$$n(x; \mu, \sigma^2) = \frac{1}{\sqrt{2\pi}\sigma} \exp\left[-\frac{(x-\mu)^2}{2\sigma^2}\right]. \tag{4.17}$$

If the reader is not completely familiar with this density function, then he should take it as a calculus exercise to trace its curve; he should determine the extent, asymptotes, maximum, points of inflection, and in general satisfy himself that this curve is in fact "bell shaped" and symmetric about μ. An easy integration shows that

$$E\mathbf{x} = \int_{-\infty}^{\infty} x \cdot n(x; \mu, \sigma^2)\, dx = \mu.$$

It is not as easy to prove that A, the area under the density function, is 1. We establish this as follows:

$$A^2 = \int_{-\infty}^{\infty}\int_{-\infty}^{\infty} \frac{1}{2\pi\sigma^2} \exp\left[-\frac{(x-\mu)^2 + (y-\mu)^2}{2\sigma^2}\right] dx\, dy$$

$$= \int_{r=0}^{\infty}\int_{\theta=0}^{2\pi} \frac{1}{2\pi} \exp\left(-\frac{r^2}{2}\right) r\, dr\, d\theta$$

where we have made the polar transformation

$$\frac{x-\mu}{\sigma} = r\cos\theta \quad \text{and} \quad \frac{y-\mu}{\sigma} = r\sin\theta.$$

Upon integrating, we have $A^2 = 1$, and hence, $A = 1$.

The following example illustrates an interesting manner in which the normal distribution may arise. The lateral and vertical miss-distance of a bullet fired at a target are denoted by $\mathbf{x}$ and $\mathbf{y}$. We assume (i) that $\mathbf{x}$ and $\mathbf{y}$ are independent r.v.'s having differentiable density functions and (ii) their joint density function is rotation invariant. It is a remarkable fact that these assumptions imply that $\mathbf{x}$ and $\mathbf{y}$ are normally distributed r.v.'s.

The assumptions yield the relation

$$f_{\mathbf{x}}(x) \cdot f_{\mathbf{y}}(y) = f(x, y) = g(r)$$

where $r = \sqrt{x^2 + y^2}$. Differentiating with respect to x and dividing through by the original equation we obtain

$$\frac{f_{\mathbf{x}}'(x)}{x \cdot f_{\mathbf{x}}(x)} = \frac{g'(r)}{r \cdot g(r)}.$$

The left-hand side of this equation does not involve y while the right-hand side does. It follows that this expression must be a constant, hence

$$[\log f_{\mathbf{x}}(x)]' = cx$$

$$f_{\mathbf{x}}(x) = ae^{cx^2}.$$

Since

$$\int_{-\infty}^{\infty} f_{\mathbf{x}}(x)\, dx = 1,$$

c is necessarily negative; we may write $c = -1/2\sigma^2$. $f_{\mathbf{x}}(x)$ is necessarily the normal density function $n(x; 0, \sigma^2)$. A similar argument may be applied to

$f_{\mathbf{y}}(y)$. Invariance of $f(x, y)$ under rotation implies that $\mathbf{x}$ and $\mathbf{y}$ have the same constant c; hence,

$$f(x, y) = (2\pi\sigma^2)^{-1} \exp\left(-\frac{x^2 + y^2}{2\sigma^2}\right).$$

One reason for the usefulness of the normal distribution is that, being a two-parameter family, it can be fitted to a wide variety of data and yet only a very abbreviated table is needed to look up probabilities.[2] Thus if $\mathbf{x}$ is $N(\mu, \sigma^2)$, that is, has density $n(x; \mu, \sigma^2)$, then

$$P(\mathbf{x} \le x) = \int_{-\infty}^{x} (2\pi\sigma^2)^{-1/2} \exp\left[-\frac{(v - \mu)^2}{2\sigma^2}\right] dv = N\left(\frac{x - \mu}{\sigma}\right)$$

where here and henceforth $N(x)$ indicates the $N(0, 1)$ d.f.

$$N(x) = \int_{-\infty}^{x} (2\pi)^{-1/2} \exp\left(-\frac{v^2}{2}\right) dv.$$

A small table of $N(x)$ follows; more detail can be found in the table on page 171.

The Standard Normal Distribution

x	0	0.5	1	1.5	2	2.5	3
$N(x)$	0.500	0.691	0.841	0.933	0.977	0.994	0.999

4.7 THE EXPONENTIAL AND GAMMA DISTRIBUTIONS

Recall the discussion of Section 3.3 concerning the collisions of a distinguished molecule of gas; there we calculated the probability $P_t(x)$ of x collisions within a fixed time interval of length t. Now reconsider the same problem from a different point of view; let $\mathbf{y}_x$ be the random time at which the xth collision occurs. From Equation (3.10),

$$\begin{aligned} P(t < \mathbf{y}_x \le t + \Delta t) &= P_t(x - 1) \cdot P_{\Delta t}(1) \\ &= \frac{(\rho t)^{x-1}}{(x - 1)!} e^{-\rho t} \cdot [\rho \,\Delta t + 0(\Delta t)]. \end{aligned}$$

Therefore, the density function of $\mathbf{y}_x$ is

$$\rho \frac{(\rho t)^{x-1}}{(x - 1)!} e^{-\rho t}, \quad t \ge 0. \tag{4.18}$$

[2] A more important reason is the central limit theorem, Theorem 7.11.

We remark that (4.18) is a density since the integral involved is a gamma function. A brief review of some of the properties of the gamma function, $\Gamma(\alpha)$, may be found in an appendix to this chapter.

In general, we will say that an r.v. $\mathbf{z}$ has the *gamma distribution* with parameters α and ρ if the density function of $\mathbf{z}$ is

$$\gamma(z; \alpha, \rho) = \begin{cases} 0, & z < 0 \\ \dfrac{\rho(\rho z)^{\alpha-1}}{\Gamma(\alpha)} e^{-\rho z}, & z \geq 0; \end{cases} \tag{4.19}$$

ρ and α must both be positive.

The special case $\alpha = 1$ has a name of its own.

$$\gamma(z; 1, \rho) = \begin{cases} 0, & z < 0 \\ \rho e^{-\rho z}, & z \geq 0 \end{cases}$$

is the density function of the *exponential distribution*.

The expected value of the gamma r.v. is

$$E\mathbf{z} = \int_0^\infty z\gamma(z; \alpha, \rho)\, dz = \frac{\alpha}{\rho},$$

that of the exponential is ρ^{-1}.

The gamma and exponential distributions have received considerable attention as distributions of life length. A rationale is as follows: We may imagine that periods of stress occur at random in time (in the sense of Section 3.3) during the life of a piece of equipment and that a fixed number x of stress periods will cause failure. Letting $\mathbf{N}_t$ be the random number of stress periods in time t and $\mathbf{y}_x$ the random time of occurrence of the xth period, then the probability of failure at or before time t is

$$P(\mathbf{N}_t \geq x) = P(\mathbf{y}_x \leq t) = \int_0^t \gamma(z; x, \rho)\, dz$$

where ρ is the expected rate of arrival of periods of stress. The expected length of life is x/ρ, that is, proportional to x, a measure of durability of the equipment, and inversely proportional to ρ, a measure of the severity of the environment.

The exponential distribution, when used as a model for duration of life, has a curious property. If $\mathbf{t}$ denotes the random length of life of, say, a piece of electronic equipment then for $t, s \geq 0$

$$P(\mathbf{t} > t + s | \mathbf{t} > s) = \frac{\int_{t+s}^\infty \rho e^{-\rho z}\, dz}{\int_s^\infty \rho e^{-\rho z}\, dz} = e^{-\rho t} = P(\mathbf{t} \geq t).$$

That is, if the piece of equipment has been working for s hours then the probability of survival for an additional t hours is exactly the same as the probability of survival for t hours of a new piece of equipment; in this sense no deterioration has occurred.

This property of no deterioration uniquely characterizes the exponential distribution. Let $F(t)$ be the d.f. of t and for s, $t > 0$, assume

$$P(\mathbf{t} > t + s | \mathbf{t} > s) = P(\mathbf{t} > t).$$

This is equivalent to

$$1 - F(t + s) = [1 - F(t)][1 - F(s)]$$

or

$$z(t + s) = z(t) + z(s)$$

where $z(t) = \log [1 - F(t)]$. But from the lemma of Section 4.3, the only right continuous additive functions are of the form $z(t) = ct$. Hence, $F(t) = 1 - e^{ct}$, for $t \geq 0$. Since

$$\lim_{t \to \infty} F(t) = 1,$$

we see that c is negative and $\mathbf{t}$ has an exponential distribution.

4.8 APPLICATION—SURVIVAL FUNCTIONS

As in the previous section, let $\mathbf{t}$ denote the length of life of an object but assume now that

$$F(t) = P(\mathbf{t} \leq t)$$

is an arbitrary continuous d.f. with density $f(t)$. $\bar{F}(t) = 1 - F(t)$, called the *survival function*, is frequently more convenient to work with in this context than the distribution function. $\bar{F}(t)$ is the probability of survival till time t.

A second function

$$r(t) = \frac{f(t)}{\bar{F}(t)}$$

has a useful probability interpretation. We have, for the probability of failure in the interval $(t, t + \Delta t)$ given survival till time t:

$$P(t < \mathbf{t} \leq t + \Delta t | \mathbf{t} > t) = \frac{F(t + \Delta t) - F(t)}{1 - F(t)} \doteq r(t)\, \Delta t.$$

$r(t)$ is called the *failure rate function*, or in actuarial work the force of mortality. For the exponential d.f., $r(t)$ is constant and equal to ρ; this is a manifestation of the no-deterioration property of the exponential. Shape of the failure rate function for actual populations will depend on the nature of the items being observed. For items which do deteriorate, such as human beings, $r(t)$ will increase as the population approaches old age. Current thinking for items such as automobiles and washing machines is that $r(t)$ will initially be high but decreasing during the "break-in period," that it will become constant during the long middle life of the items and ultimately become an increasing function as deterioration takes effect.

The failure rate and survival functions can be determined from one another using the formulas

$$r(t) = \frac{f(t)}{\bar{F}(t)}$$

and

$$\bar{F}(t) = \exp\left[-\int_0^t r(t)\,dt\right].$$

Common survival distribution models are the exponential, the gamma, the *Weibull*

$$f(t) = \rho\alpha t^{\alpha-1}e^{-\rho t^{\alpha}}; \quad \rho, \alpha > 0; \quad t \geq 0$$

with

$$r(t) = \rho\alpha t^{\alpha-1}$$

and the *truncated normal*

$$\begin{aligned} f(t) &= \frac{k}{\sigma\sqrt{2\pi}}\exp\left[-\frac{(t-\mu)^2}{2\sigma^2}\right]; & t > 0 \\ &= 0; & t \leq 0. \end{aligned}$$

4.9 APPLICATION—CAR FOLLOWING

The Poisson distribution is the main theoretical instrument used to model the distribution of automobile traffic on a highway. The probability of x arrivals in t seconds with traffic flow q is

$$p(x; qt) = \frac{(qt)^x e^{-qt}}{x!}, \quad x = 0, 1, 2, \ldots.$$

Defining headway $\mathbf{h}$ as the time gap between cars then

$$P(\mathbf{h} > t) = p(0; qt) = e^{-qt}$$

and the d.f. of headway is

$$P(\mathbf{h} \leq t) = 1 - e^{-t/E\mathbf{h}}$$

where $q = (E\mathbf{h})^{-1}$. The relation between the Poisson counting distribution and the exponential headway distribution is important in practice, since it is easier to count vehicles passing an observer than it is to measure time gaps between cars.

The results of the above equations agree with actual observation for low density traffic but they break down under congested conditions for two reasons. First, automobiles are not points; they have length and they must follow one another at some minimum safe distance. Second, they cannot pass at will.

The second difficulty is substantial and we do not discuss it here (however, refer to Section 10.4). The first problem can be partially overcome in two ways. First, the exponential curve can be shifted to the right by an amount τ equal to the minimum headway. Thus,

$$P(\mathbf{h} \le t) = 1 - \exp\left[-\frac{(t - \tau)}{E\mathbf{h}}\right], \quad t \ge \tau.$$

Second, gaps less than the minimum safe headway can be modeled as unlikely rather than impossible by assuming a gamma distribution for headway

$$P(\mathbf{h} \le t) = \int_0^t \frac{q^\alpha}{\Gamma(\alpha)} z^{\alpha-1} e^{-qz}\, dz.$$

4.10 THE BETA DISTRIBUTION

Here we show that an addition theorem holds for gamma distributions and we introduce the important *beta distribution* with density

$$\beta(w; \mu, \nu) = \frac{\Gamma(\mu + \nu)}{\Gamma(\mu)\cdot\Gamma(\nu)} w^{\mu-1}(1 - w)^{\nu-1} \tag{4.20}$$

where $\mu, \nu > 0$ and $0 \le w \le 1$.

Theorem 4.2

If $\mathbf{x}$ and $\mathbf{y}$ are independent r.v.'s having densities $\gamma(x; \mu, \rho)$ and $\gamma(y; \nu, \rho)$ then $\mathbf{r} = \mathbf{x} + \mathbf{y}$ and $\mathbf{w} = \mathbf{x}/(\mathbf{x} + \mathbf{y})$ have the densities $\gamma(r; \mu + \nu, \rho)$ and $\beta(w; \mu, \nu)$.

Proof

The joint density function of $\mathbf{x}$ and $\mathbf{y}$ is

$$\frac{\rho^{\mu+\nu}}{\Gamma(\mu)\Gamma(\nu)} x^{\mu-1} y^{\nu-1} e^{-\rho(x+y)}; \quad x, y \ge 0.$$

The single-valued transformation $r = x + y$, $w = x/(x + y)$ or $y = r(1 - w)$, $x = r \cdot w$ has Jacobian r, so that **w** and **r** have joint density

$$\frac{\Gamma(\mu + \nu)}{\Gamma(\mu) \cdot \Gamma(\nu)} w^{\mu - 1}(1 - w)^{\nu - 1} \cdot \gamma(r; \mu + \nu, \rho).$$

The marginal density of each r.v. is obtained by integrating out the other.

Many of the common probability functions can be obtained from the beta distribution. An example is the important identity

$$\sum_{x=k}^{n} \binom{n}{x} p^x (1 - p)^{n-x} = \frac{\Gamma(n + 1)}{\Gamma(k)\Gamma(n - k + 1)} \int_0^p w^{k-1}(1 - w)^{n-k}\, dw \tag{4.21}$$

which is valid for k and n integral and $0 \leq p \leq 1$. Stated in probabilistic terms this identity is equivalent to the following theorem.

Theorem 4.3

If **x** has the binomial probability function $b(x; n, p)$ while **w** has beta density $\beta(w; k, n - k + 1)$, then $P(\mathbf{x} \geq k) = P(\mathbf{w} \leq p)$.

Proof

Integrating the right side of (4.21) by parts we have

$$\int_0^p w^{k-1}(1 - w)^{n-k}\, dw = \frac{p^k(1 - p)^{n-k}}{k} + \frac{n - k}{k} \int_0^p w^k (1 - w)^{n-k-1}\, dw$$

or

$$\int_0^p \beta(w; k, n - k + 1)\, dw = b(k; n, p) + \int_0^p \beta(w; k + 1, n - k)\, dw.$$

Applying this recursion relationship $n - k$ times yields (4.21).

4.11 MORE ON CONDITIONAL PROBABILITY

The elementary definition of $P(B|C)$ leaves conditional probabilities undefined when $P(C) = 0$. This causes no difficulty for discrete r.v.'s but the definition needs to be extended so that we may define conditional density functions. Our method of extension will be to consider a limiting operation. Roughly, we consider a sequence of monotone decreasing sets $X_1, X_2, \ldots$ all having nonzero probability and such that $\bigcap X_i = C$ and we define

$$P(B|C) = \lim_{i \to \infty} \frac{P(B \cap X_i)}{P(X_i)}.$$

A general concept of limit will prove useful for this purpose. To present this purely mathematical concept, we momentarily depart from probability proper. A real-valued set function $f(X)$ with domain $\mathbf{C}$ and a family $\mathbf{D}$ of sets each contained in $\mathbf{C}$ is given. The family $\mathbf{D}$, called a direction, has the properties:

(i) $\mathbf{D}$ is not empty; it contains at least one set A.
(ii) If A_1 and A_2 are in $\mathbf{D}$, then there is A_3 in $\mathbf{D}$ such that $A_3 \subset A_1 \cap A_2$.

We write

$$\lim_{\mathbf{D}} f(X) = k$$

to mean that if N is a neighborhood of k then there exists A in $\mathbf{D}$ such that $f(X) \in N$ for all X in $\mathbf{C}$ which are subsets of A.

With this machinery we may prove most of the standard theorems about limits. We particularly need two results.

Theorem 4.4

If $\lim_{\mathbf{D}} f(X) = k$ and $\lim_{\mathbf{D}} f(X) = h$, then $h = k$.

Proof

Assume two limits h and k with $h < k$; pick a point c between them. $(-\infty, c)$ is a neighborhood of h so that there must exist $A_1 \in \mathbf{D}$ such that for all $X \in \mathbf{C}$ satisfying $A \supset X, f(X) < c$. Likewise, since (c, ∞) is a neighborhood of k, there must exist $A_2 \in \mathbf{D}$ such that for all $X \in \mathbf{C}$ satisfying $A_2 \supset X$, $f(X) > c$. By property (ii) there is $A_3 \in \mathbf{D}$ such that $A_3 \subset A_1 \cap A_2$. Now, $f(A_3) < c$ and $f(A_3) > c$, a contradiction.

Theorem 4.5

If $f(X)$ and $g(X)$ have common domain $\mathbf{C}$,

$$\lim_{\mathbf{D}} f(X) = k, \qquad \lim_{\mathbf{D}} g(X) = h \neq 0,$$

and the quotients $f(X)/g(X)$ are defined for X in $\mathbf{C}$ then

$$\lim_{\mathbf{D}} \frac{f(X)}{g(X)} = \frac{k}{h}.$$

The proof of this theorem follows standard lines. For every ϵ there will be A_1 such that $k - \epsilon < f(X) < k + \epsilon$ whenever $X \subset A_1$, and A_2 such that $h - \epsilon < g(X) < h + \epsilon$ whenever $X \subset A_2$. From property (ii) there is $A_3 \subset A_1 \cap A_2$ such that

$$\frac{k - \epsilon}{h + \epsilon} < \frac{f(X)}{g(X)} < \frac{k + \epsilon}{h - \epsilon}$$

whenever $X \subset A_3$. Since ϵ is arbitrary, this pair of inequalities proves the theorem.

Now consider events B and C in a probability space. Take

$$\mathbf{C} = \{X: P(X) > 0\}$$

and $\mathbf{D}$ a direction whose sets contain C. Given the direction $\mathbf{D}$, define

$$P(B|C) = \lim_{\mathbf{D}} \frac{P(B \cap X)}{P(X)}.$$

In this context, Theorem 4.4 now says that $P(B|C)$ is uniquely defined. Theorem 4.5 says that if $P(C) > 0$, then $P(B|C) = P(B \cap C)/P(C)$; that is, this new concept of conditional probability generalizes the old.

Conditional probabilities defined in this way satisfy Kolmogorov's axioms, the certain event now being C. That is, $P(C|C) = 1$, $P(B|C) \geq 0$ for all B and $P(B_1 + B_2 + \cdots|C) = P(B_1|C) + P(B_2|C) + \cdots$, for pairwise disjoint events $B_1, B_2, \ldots$. Hence, conditional probabilities are one kind of axiomatic probability and all probability theorems and concepts have their conditional probability counterpart.

In particular, the conditional expectation of an r.v. $\mathbf{x}$ given that the event C has occurred need not be defined anew.

We now specialize the general definition to various cases. If $p_{\mathbf{y}}(y) > 0$, then from Theorem 4.5 $p_{\mathbf{x}|\mathbf{y}}(x, y)$, the *conditional probability function* of $\mathbf{x}$ given that $\mathbf{y} = y$, is given by

$$p_{\mathbf{x}|\mathbf{y}}(x, y) = \frac{P(\mathbf{x} = x, \mathbf{y} = y)}{P(\mathbf{y} = y)} = \frac{p(x, y)}{p_{\mathbf{y}}(y)}.$$

For r.v.'s having density function $f(x, y)$ with marginal density $f_{\mathbf{y}}(y) > 0$, the collection of all sets of the form $X = \{(t, u): y - h < u \leq y\}$ constitute a direction with $P(X) > 0$. We define the *conditional d.f.* of $\mathbf{x}$ given that $\mathbf{y} = y$ as

$$\begin{aligned} F_{\mathbf{x}|\mathbf{y}}(x, y) &= \lim_{h \to 0} P(\mathbf{x} \leq x | y - h < \mathbf{y} \leq y) \\ &= \lim_{h \to 0} \frac{\int_{-\infty}^{x} \int_{y-h}^{y} f(t, u)\, du\, dt}{\int_{y-h}^{y} f_{\mathbf{y}}(u)\, du} \\ &= \int_{-\infty}^{x} \frac{f(t, y)}{f_{\mathbf{y}}(y)}\, dt. \end{aligned}$$

This equation in turn leads us to define the *conditional density function*

$$f_{\mathbf{x}|\mathbf{y}}(x, y) = \frac{\partial F_{\mathbf{x}|\mathbf{y}}(x, y)}{\partial x} = \frac{f(x, y)}{f_{\mathbf{y}}(y)}$$

for $f_{\mathbf{y}}(y) > 0$. The conditional expectation $E(\mathbf{x}|\mathbf{y} = y)$ is given by

$$E(\mathbf{x}|\mathbf{y} = y) = \int_{-\infty}^{\infty} x f_{\mathbf{x}|\mathbf{y}}(x, y)\, dx = \frac{1}{f_{\mathbf{y}}(y)} \int_{-\infty}^{\infty} x f(x, y)\, dx.$$

We have already said that two r.v.'s are called *independent* if

$$P[(\mathbf{x} \in A) \cap (\mathbf{y} \in B)] = P(\mathbf{x} \in A) \cdot P(\mathbf{y} \in B)$$

for all sets A and B in the ranges of $\mathbf{x}$ and $\mathbf{y}$, respectively, and that this is equivalent to requiring that the d.f.'s multiply: $F(x, y) = F_{\mathbf{x}}(x) \cdot F_{\mathbf{y}}(y)$. Continuous r.v.'s will then be independent if their density functions multiply, that is, $f(x, y) = f_{\mathbf{x}}(x) \cdot f_{\mathbf{y}}(y)$, or equivalently (for $f_{\mathbf{y}}(y) > 0$) if $f_{\mathbf{x}|\mathbf{y}}(x, y) = f_{\mathbf{x}}(x)$. A finite number of r.v.'s are called independent if the multiplication formula holds for every subset of the variables.

PROBLEMS

1. Prove the following "addition theorem" for normal independent r.v.'s. If $\mathbf{x}$ and $\mathbf{y}$ have densities $n(x; \mu_x, \sigma_x^2)$ and $n(y; \mu_y, \sigma_y^2)$, then $\mathbf{z} = \mathbf{x} + \mathbf{y}$ has density $n(z; \mu_x + \mu_y, \sigma_x^2 + \sigma_y^2)$.

2. Let $y(\ell)$ be the probability that a string of length ℓ does not break under a constant load and assume $y(\ell_1 + \ell_2) = y(\ell_1) \cdot y(\ell_2)$. Calculate the expected breaking length of the string.

3. Let $\mathbf{x}_1, \mathbf{x}_2, \ldots,$ be a sequence of independent r.v.'s each having the d.f. $F(x)$, that is,

$$P(\mathbf{x}_i \leq x) = F(x) \qquad \text{for } i = 1, 2, \ldots.$$

Let $\mathbf{N}(x)$ be the number of these r.v.'s observed before obtaining a value $> x$. Show that $E\mathbf{N}(x) = [1 - F(x)]^{-1}$.

4. Let $\mathbf{x}$ and $\mathbf{y}$ have the joint density $(2\pi)^{-1} \exp[-\frac{1}{2}(x^2 + y^2)]$. Derive the joint density of $\mathbf{r}$ and $\mathbf{a}$ where $\mathbf{x} = \mathbf{r} \cos \mathbf{a}$ and $\mathbf{y} = \mathbf{r} \sin \mathbf{a}$.

5. Buses scheduled for arrival at times 0 and 1 arrive instead at times $\mathbf{x}_0$ and $1 + \mathbf{x}_1$ $(0 \leq \mathbf{x}_i \leq 1)$. $\mathbf{x}_0$ and $\mathbf{x}_1$ are independent and identically distributed with density $f(x)$. What is the expected waiting time of a person arriving at random in the interval [0, 1]?

6. Show that

$$d(x; \alpha, \beta, \gamma) = K^{-1} \exp\left[-\frac{1}{2}\left|\frac{x - \alpha}{\gamma}\right|^{1/\beta}\right], \qquad -\infty < x < \infty \tag{4.22}$$

is a well-defined density for $-\infty < \alpha < \infty$, $0 < \gamma$, $0 < \beta$, and that $K = \gamma \cdot 2^{\beta+1} \cdot \Gamma(\beta + 1)$. The special case $\beta = 1$ is the distribution of *Laplace*. Note that $\beta = \frac{1}{2}$ yields the normal and that

$$\lim_{\beta \to 0} d(x; \alpha, \beta, \gamma) = u(x; \alpha - \gamma, \alpha + \gamma),$$

the uniform density.

7. If the r.v. $\mathbf{r}$ has the density function

$$f(x) = \begin{cases} R^{-1}e^{-x/R}, & x \geq 0 \\ 0, & x < 0 \end{cases}$$

then what is the density function of $\mathbf{s} = k \log \mathbf{r}$?

8. What is the density function of $\mathbf{y} = \mathbf{x}^2$ in terms of the density of $\mathbf{x}$? What is the density of $\mathbf{x}$ in terms of that of $\mathbf{y}$?

9. If $\mathbf{x}$ and $\mathbf{y}$ are independent random variables each having the exponential density function, e^{-x} for $x \geq 0$, then what is the distribution of $\mathbf{s} = \mathbf{x} + \mathbf{y}$?

10. (a) An event E may or may not occur in any interval of time. If $p = P$ (E occurs in time Δt) and $q = 1 - p = P$ (E does not occur in time Δt) then, assuming independence over disjoint intervals, show that E occurs for the first time in the $k + 1$th ($k = 0, 1, 2, \ldots$) interval with probability $q^k p$.
    (b) Writing $p = \lambda \cdot \Delta t$ in part (a) obtain a limiting expression for the probability that the first success occurs in the interval $(t, t + \Delta t)$.
    (c) Show that your answer to (b) reflects a true probability distribution.

11. If $E(\mathbf{y}|\mathbf{x} = x) = E\mathbf{y}$ for all x then $\mathbf{x}$ and $\mathbf{y}$ are uncorrelated. Show this in the special case where $\mathbf{x}$ and $\mathbf{y}$ have a density function.

12. Produce an example of uncorrelated r.v.'s (not necessarily having a density) for which the formula $E(\mathbf{y}|\mathbf{x} = x) = E\mathbf{y}$ does not hold.

13. Show that the definitions of conditional probability function and conditional d.f. are consistent by proving that if $p_{\mathbf{y}}(y) > 0$, the jump in $F_{\mathbf{x}/\mathbf{y}}(x, y)$ at the point x_0 is $p_{\mathbf{x}/\mathbf{y}}(x_0, y)$.

14. If $\mathbf{x}$ and $\mathbf{y}$ are independent exponential r.v.'s with expectations λ_1^{-1} and λ_2^{-1}, show (i) $P(\mathbf{x} > \mathbf{y}) = \lambda_2/(\lambda_1 + \lambda_2)$ and (ii) min $(\mathbf{x}, \mathbf{y})$ is exponential with expectation $(\lambda_1 + \lambda_2)^{-1}$.

15. (Buffon's needle problem). A large plane area is ruled with equidistant parallel straight lines and a needle is thrown down. What is the probability that the needle touches one of the lines?

16. For $n = 1, 2, \ldots$, show that

$$\int_0^t \gamma(x; n + 1, \rho)\, dx = -\frac{(\rho t)^n e^{-\rho t}}{n!} + \int_0^t \gamma(x; n, \rho)\, dx.$$

Provide an interpretation similar to that of Theorem 4.3.

17. Let $\mathbf{x}_1, \mathbf{x}_2, \ldots, \mathbf{x}_n$ be independent r.v.'s all having the same d.f. $F(x)$. Let $\mathbf{y}$ denote the number of these r.v.'s which do not exceed x. Prove that

$$P(\mathbf{y} = k) = \binom{n}{k}[F(x)]^k[1 - F(x)]^{n-k}, \qquad k = 0, 1, \ldots, n.$$

18. Find the density corresponding to the d.f.

$$F(x) = \begin{cases} 0 & \text{if } x < 0 \\ (2/\pi) \text{ Arc sin } x & \text{if } 0 \leq x < 1 \\ 1 & \text{if } x \geq 1. \end{cases}$$

19. A number $\mathbf{x}$ is chosen at random from among the integers, 1, 2, 3, and 4. Another, $\mathbf{y}$, is chosen from among those which are at least as large as $\mathbf{x}$.
    (a) Find the marginal probability function of $\mathbf{x}$ and the conditional probability function of $\mathbf{y}$ for each value of $\mathbf{x}$.
    (b) Determine the joint probability function of $\mathbf{x}$ and $\mathbf{y}$ and marginal probability function of $\mathbf{y}$.
    (c) Determine the marginal probability function of $\mathbf{y}$ directly from the conditional probability function of $\mathbf{y}$ and the marginal probability function of $\mathbf{x}$.
    (d) Are $\mathbf{x}$ and $\mathbf{y}$ independent? Why?

20. Given the conditional probability density

$$g_{\mathbf{y}|\mathbf{x}}(y, x) = \begin{cases} 8xy, & \text{for } 0 < y < x \\ 0, & \text{for } y < 0 \quad \text{or} \quad y > x \end{cases}$$

and the density function of $\mathbf{x}$, $f(x) = 4x^3$ where $0 < x < 1$, find
    (a) The joint density function of $\mathbf{x}$, $\mathbf{y}$;
    (b) The marginal density function of $\mathbf{y}$;
    (c) $P(\mathbf{x} > 2\mathbf{y})$.

21. Let $f(x, y) = ky$ for $0 < x < 1$, $0 < y < 1$ and $f(x, y) = 0$ otherwise.
    (a) Determine k so that $f(x, y)$ is a joint density.
    (b) Find $f\mathbf{y}(y)$ and $F_{\mathbf{y}}(y)$.
    (c) Find $f_{\mathbf{x}|\mathbf{y}}(x, y)$.

22. Given $f(x, y) = xe^{-x(y+1)}$, $x \geq 0$, $y \geq 0$, find
    (a) The marginal probability density functions of $\mathbf{x}$ and $\mathbf{y}$.
    (b) The conditional probability density functions of $\mathbf{x}$ given $\mathbf{y}$ and of $\mathbf{y}$ given $\mathbf{x}$.

23. Consider two r.v.'s $\mathbf{x}$ and $\mathbf{y}$. The conditional distribution of $\mathbf{x}$ given $\mathbf{y} = y$ is Poisson with parameter y; the marginal distribution of $\mathbf{y}$ is gamma. What is the marginal distribution of $\mathbf{x}$?

24. The r.v. $\mathbf{x}$ has continuous d.f. $F(x)$. Find the d.f. of (a) $\mathbf{y} = F(\mathbf{x})$ and (b) $\mathbf{z} = -\ln F(\mathbf{x})$.

25. Consider the density function $f_{\mathbf{x}}(x) = x/4$ for $1 < x < 3$. Obtain the density of $\mathbf{y} = (1.5 - \mathbf{x})^2$.

26. The joint density of **x** and **y** is

$$f(x, y) = \frac{1}{4\pi^2}[1 - \sin(x + y)], \qquad -\pi \le x, y \le \pi$$
$$= 0, \qquad \text{otherwise.}$$

(a) Obtain the marginal d.f. of **x**.
(b) Are **x** and **y** independent? Why?

27. Show that the points of discontinuity of a d.f. are countable.

APPENDIX—THE GAMMA FUNCTION

If the function $g(t)$ of (4.18) is to be a density, then we must have

$$\int_0^\infty g(t)\,dt = 1$$

which is the same as

$$\int_0^\infty y^{x-1}e^{-y}\,dy = (x - 1)!$$

Hence the integral

$$\int_0^\infty y^{x-1}e^{-y}\,dy \tag{4.23}$$

must at least converge for positive integral values of x. The following is a more general result:

Theorem

The integral (4.23) converges or diverges according as $x > 0$ or $x \le 0$. For $x > 0$, the integral (4.23) is called the *gamma function* and denoted by $\Gamma(x)$.

Proof

Case 1, $x > 0$. From l'Hospital's rule, $y^{x+1}e^{-y} \to 0$ as $y \to \infty$. Thus for every ϵ, there exists Y such that $y^{x+1}e^{-y} < \epsilon$ whenever $y > Y$. For $b > Y$,

$$\int_0^b y^{x-1}e^{-y}\,dy \le \int_0^Y y^{x-1}\,dy + \int_Y^b \frac{\epsilon}{y^2}\,dy = \frac{Y^x}{x} + \epsilon\left(\frac{1}{Y} - \frac{1}{b}\right)$$
$$\le \frac{Y^x}{x} + \frac{\epsilon}{Y}.$$

Therefore,

$$\lim_{b\to\infty}\int_0^b y^{x-1}e^{-y}\,dy \le \frac{Y^x}{x} + \frac{\epsilon}{Y}$$

and the integral $\Gamma(x)$ exists.

Case 2, $x \le 0$. For $y < 1$, $\log y < 0$ and $x \log y \ge 0$ so that $y^x \ge 1$ and $y^{x-1}e^{-y} \ge (ey)^{-1}$. Now

$$\int_a^\infty y^{x-1}e^{-y}\,dy \ge \int_a^1 y^{x-1}e^{-y}\,dy \ge \int_a^1 (ey)^{-1}\,dy = \frac{-\log a}{e}.$$

But $-\log a \to \infty$ as $a \to 0$ so the integral on the left cannot have a finite limit.

Other useful facts concerning the gamma function $\Gamma(x)$ are: (i) $\Gamma(x) = (x-1)\cdot\Gamma(x-1)$, (ii) $\Gamma(x) = (x-1)!$ for positive integral x, and (iii) $\Gamma(\frac{1}{2}) = \sqrt{\pi}$. The proof of (i) involves an integration by parts and (ii) follows from (i). Statement (iii) was essentially proved in connection with the normal distribution. The connection is as follows:

$$\begin{aligned}\Gamma(\tfrac{1}{2}) &= \int_0^\infty y^{-1/2}e^{-y}\,dy = \sqrt{2}\int_0^\infty e^{-w^2/2}\,dw \\ &= \frac{1}{\sqrt{2}}\int_{-\infty}^\infty e^{-w^2/2}\,dw = \sqrt{\pi}.\end{aligned}$$

5

*APPLICATION—THE KINETIC THEORY OF AN IDEAL GAS

5.1 MAXWELL'S DISTRIBUTION

The kinetic theory as we know it today has its origin in a paper entitled "Concerning the Nature of the Motion which We Call Heat." This paper was published in 1857 by Rudolf Julius Emmanuel Clausius to whom, according to Maxwell, we owe the first complete dynamical theory of gases. In 1860, inspired by Clausius' research, James Clerk Maxwell published in the *Philosophical Magazine* an article entitled "Illustrations of the Dynamical Theory of Gases." In that paper, there appeared for the first time a derivation of the stable state velocity distribution for gas molecules which has since come to be known as Maxwell's distribution. The following discussion follows Maxwell's original derivation where this is feasible.

Throughout, the gas particles are treated as a very large number of perfectly elastic spherical balls which are moving about in a perfectly elastic vessel. In order for the mathematics to describe the physical phenomenon, it is sufficient that the gas particles be centers of force which *act like* perfectly elastic rebounding balls. After a large number of collisions, it is to be expected that the total momentum will be shared among the particles according to some regular law, the proportion of particles whose velocities lie between certain limits being stable, though the velocity of each particle changes at every collision.

Let **u**, **v**, and **w** be the components of velocity of an arbitrary particle in three rectangular directions and let $g(u, v, w)$ be their joint density function. The essence of Maxwell's derivation is that (1) **u**, **v**, and **w** are taken to be independent random variables,[1] and (2) the density function g is assumed to

[1] In 1866 Maxwell published a second derivation of his distribution, commenting that the assumption of probabilistic independence of the three velocity components "may appear precarious."

depend on the velocity magnitude but not on its direction. The second of these assumptions is reasonable because of the arbitrary nature of the underlying rectangular coordinate system, but the first would seem to require a keen physical intuition. In any case, the two assumptions taken together imply that if $f(u)$ is the (necessarily) common density of the three velocity components, then $g(u, v, w) = f(u)f(v)f(w) = \Phi(r)$, say, with $\mathbf{r} = (\mathbf{u}^2 + \mathbf{v}^2 + \mathbf{w}^2)^{1/2}$. As in the two-dimensional case (see Section 4.6), we may establish under weak regularity assumptions that the only solution of this functional equation is

$$f(u) = (\alpha^2\pi)^{-1/2} \exp(-u^2/\alpha^2), \qquad \Phi(r) = (\alpha^2\pi)^{-3/2} \exp(-r^2/\alpha^2).$$

From this result we draw the following conclusions:

(i) The proportion of particles whose velocity, resolved in a certain direction, lies between u and $u + du$ is [2]

$$\left(\frac{1}{\alpha\sqrt{\pi}}\right) \exp\left(-\frac{u^2}{\alpha^2}\right) du. \tag{5.1}$$

That is, $\mathbf{u}$ has density $n(u; 0, \alpha^2/2)$.

(ii) The proportion with velocity magnitude (speed) between r and $r + dr$ is

$$\left(\frac{4r^2}{\alpha^3\sqrt{\pi}}\right) \exp\left(-\frac{r^2}{\alpha^2}\right) dr. \tag{5.2}$$

This is *Maxwell's distribution.*

(iii) The mean speed is

$$E(\mathbf{r}) = \frac{2\alpha}{\sqrt{\pi}}\cdot \tag{5.3}$$

(iv) The mean square speed is

$$E(\mathbf{r}^2) = \tfrac{3}{2}\alpha^2. \tag{5.4}$$

5.2 CALCULATION OF THE PRESSURE OF A GAS ACCORDING TO THE KINETIC THEORY

We consider the impact of the many moving molecules of a gas upon an area A of the boundary of the container. Let the components of velocity of any molecule be taken with respect to some fixed rectangular coordinate system in space. Assume the coordinate system to be chosen so that A is perpendicular to the direction of the component $\mathbf{u}$. The number of molecules

[2] For historical reasons we have decided to preserve Maxwell's use of infinitesimal quantities. Note that Maxwell is employing the frequency interpretation of probability.

with a given **u** component u, which strike the area A in a small interval of time dt, is equal to the number that lie within a distance $u \cdot dt$ from A at the beginning of this interval. Thus, if n is the number of molecules in unit volume then according to Maxwell's conclusion (i), the total number of particles whose velocity lies between u and $u + du$ and which impinge upon the area A is

$$nAu\, dt\left(\frac{1}{\alpha\sqrt{\pi}}\right) \exp\left(-\frac{u^2}{\alpha^2}\right) du, \quad 0 < u < \infty.$$

Each molecule (with mass m) that strikes the area transfers 2 $m\mathbf{u}$ units of momentum, half of which occurs in impact and the other half during rebound. Hence, the total transfer of momentum to A in time dt is

$$\begin{aligned} T &= \int_{u=0}^{\infty} 2mu \cdot nAu\, dt \cdot \left(\frac{1}{\alpha\sqrt{\pi}}\right) \exp\left(-\frac{u^2}{\alpha^2}\right) du \\ &= 2mnA\, dt\,(\tfrac{1}{2}E\mathbf{u}^2) = \tfrac{1}{2}mnA\, dt\alpha^2. \end{aligned}$$

Now, mn = total mass per unit volume = density = ρ(say), and from Maxwell's conclusion (iv), $\frac{3}{2}\alpha^2$ = mean value of $\mathbf{r}^2 = E\mathbf{r}^2$. Thus

$$T = \tfrac{1}{3}\rho E\mathbf{r}^2 \cdot A\, dt.$$

But $T = pA\, dt$, where p is the pressure (force per unit area) on the containing vessel. Hence, we have finally

$$p = \tfrac{1}{3}\rho E\mathbf{r}^2. \tag{5.5}$$

As is expressly implied by the title of Clausius' original paper on the kinetic theory (see page 79), the mean square velocity of a gas depends only on the temperature. Accepting this fact, Equation (5.5) yields Boyle's experimentally determined law that pressure and density are proportional at constant temperature.

5.3 INTERACTING SYSTEMS OF PARTICLES

Next consider two systems of particles each moving independently according to the laws of Section 5.1. What is the proportion of pairs of particles, one of each system, whose relative velocity lies between given limits? Considering the velocity components of the two systems in a particular direction, we see that the relative velocity component has the distribution of the difference of two independent variates each having the frequency function (5.1). But the difference of two normal deviates is normal, and hence, if α_1 and α_2 are the parameters of the two systems then relative velocity in a particular direction has the frequency function (5.1) with α^2 replaced by $\alpha_1^2 + \alpha_2^2$. With this same replacement, Equations (5.2), (5.3), and (5.4) likewise hold for relative velocities between pairs of particles in the two systems.

5.4 THE MEAN FREE PATH OF A PARTICLE

As a preliminary to taking up the main topic of this section we ask: If a given particle moves with velocity y in a system behaving according to the law of Section 5.1, then what is the proportion of particles which have speed (relative to the given particle) between a and $a + da$?

The probability density of particles at the velocity point (u, v, w) is

$$\frac{1}{\alpha^3\pi^{3/2}}\, e^{-(u^2+v^2+w^2)/\alpha^2}.$$

From this we have to deduce the proportion of particles in a spherical shell of radius a and thickness da with center at distance y from the origin. We may as well assume the velocity y to be parallel to the component u. The required density of particles is then given by

$$f(a) = \iint_S \frac{1}{\alpha^3\pi^{3/2}}\, e^{-[(u-y)^2+v^2+w^2]/\alpha^2}\, d\sigma$$

where $d\sigma$ is the element of area over the surface S of the required spherical shell. Introducing polar coordinates $u = a \cos \Theta$, $v = a \sin \Theta \cos \Phi$, and $w = a \sin \Theta \sin \Phi$ then

$$f(a) = \int_0^\pi \frac{a^2 \sin \Theta}{\alpha^3\pi^{3/2}}\, e^{-[(y-a\cos\Theta)^2 + a^2\sin^2\Theta]/\alpha^2}\, d\Theta \int_0^{2\pi} d\Phi$$

$$= \frac{a}{\alpha y\sqrt{\pi}}\left(e^{-(y-a)^2/\alpha^2} - e^{-(y+a)^2/\alpha^2}\right). \tag{5.6}$$

Integrating over the range of a, we have, as a corollary, the mathematical result

$$\int_0^\infty a\left(e^{-(y-a)^2/\alpha^2} - e^{-(y+a)^2/\alpha^2}\right) da = \alpha y\sqrt{\pi}. \tag{5.7}$$

One further preliminary result is needed before proceeding to the main business of this section. If a particle moves with velocity y relative to a number of particles of which there are N in unit of volume, then the expected number of these which it approaches within a distance s in unit of time is the expected number contained within a cylinder having radius s, height equal to the magnitude of y, and the path of the particle as principal axis. This number is

$$N\pi y s^2. \tag{5.8}$$

Now, Maxwell finds the mean free path by considering two sets of particles which move as in Section 5.3; we find the expected number of pairs which approach within a distance s in unit of time. If N_2 is the number of particles of the second kind in unit volume then the number of these which have a velocity magnitude between r and $r + dr$ is, from Maxwell's conclusion (ii),

$$n_2 = N_2 \frac{4r^2}{\alpha_2{}^3\sqrt{\pi}} e^{-r^2/\alpha_2{}^2}\, dr.$$

From Equation (5.6) the number of the first kind whose speed relative to these is between a and $a + da$ is

$$n_1 = N_1 \frac{a}{\alpha_1 r\sqrt{\pi}} (e^{-(r-a)^2/\alpha_1{}^2} - e^{-(r+a)^2/\alpha_1{}^2})\, da$$

where N_1 is the number of particles in unit volume. From Equation (5.8), the expected number of pairs which approach within distance s in unit of time is

$$n_1 n_2 \pi a s^2 = N_1 N_2 \frac{4}{\alpha_1 \alpha_2{}^3} s^2 a^2 r e^{-r^2/\alpha_2{}^2}\{e^{-(r-a)^2/\alpha_1{}^2} - e^{-(r+a)^2/\alpha_1{}^2}\}\, da\, dr.$$

Equation (5.7) allows us to integrate with respect to r, and integrating again over the range of a, yields

$$2N_1 N_2 \sqrt{\pi}\sqrt{\alpha_1{}^2 + \alpha_2{}^2}\, s^2$$

for the expected number of collisions in unit time within unit volume between particles of different kind, s being the distance between centers at collision. The number of collisions between particles of the first kind, s_1 being the striking distance, is

$$2N^2\sqrt{\pi}\sqrt{2\alpha_1{}^2}\, s_1{}^2$$

and the mean speed of this system is $2\alpha_1/\sqrt{\pi}$ so that if ℓ_1 is the mean distance traveled between collisions and k is the mean number of collisions per particle per unit of time then $k \cdot \ell_1 =$ mean speed and

$$\frac{1}{\ell_1} = \pi N_1 \sqrt{2} s_1{}^2 + \pi N_2 \frac{\sqrt{\alpha_1{}^2 + \alpha_2{}^2}}{\alpha_1} s^2.$$

If the particles of the two systems are of the same kind and ℓ is their common *mean free path* then

$$\ell = (\pi\sqrt{2}s^2N)^{-1}$$

where $N = N_1 + N_2$ is the total number of particles present.

To find the probability of a particle reaching a given distance before striking any other, let us suppose that the probability of a particle being stopped while passing through a distance dx is $\beta\, dx$. Then according to Equation (4.18) the density function of distance traveled between collisions is $\beta e^{-\beta x}$. To determine β, note that the mean of this distribution, which is $1/\beta$, is also the mean free path; hence,

$$\beta = \pi\sqrt{2}s^2N.$$

PROBLEM

1. By choosing a random direction in space we mean choosing a point on the unit sphere, the probability of any region of the sphere being equal to the area of the region divided by the area of the sphere. Show that if velocity, direction, and magnitude are independent and random, then **u**, **v**, **w**, the rectangular components of velocity, are independent and identically distributed.

6

*APPLICATION—LINEAR PAIRED COMPARISON MODELS

6.1 PAIRED COMPARISONS

Chapter 5 treats in some detail an important physical science application of probability. We now wish to provide one slightly involved example of probabilistic reasoning which is more typical of the behavioral sciences.

In many experimental contexts in the behavioral sciences preference relations among objects can be obtained where numerical measurement is difficult or impossible. This situation occurs, for example, in regard to individual preferences for pieces of music. Frequently because of memory, fatigue, or distance limitations items will be presented to a subject only in pairs and he will be asked to state his preference. In Section 6.3 we will treat a class of models for such paired comparison experiments.

Much of our treatment uses the traditional psychophysical ideas of stimulus and sensation. Stimulus is a measurable physical quantity (like sound measured in decibels). The stimuli being considered are denoted collectively by X and individually by x. We treat X as a portion of the real line; x is then a real number.

Paired comparison experiments may be formulated in terms of stimulus as follows: A subject, exposed to two stimuli of the same kind, reports which he sees as the larger. The paired comparison experiment is clearly probabilistic in nature since a subject will sometimes report the larger stimulus as smaller.

Sensation is a construct introduced to explain behavior. Presumably, two stimuli will have a definite degree of psychological similarity which can be operationally defined in terms of the frequency with which they are confused in a paired comparison task. Let $\pi(x_1, x_2)$ be the frequency probability that a subject reports a stimulus x_1 as greater than x_2.

The method of paired comparisons is now attributed to Fechner[1] who used the method in his investigations of Weber's law.

6.2 THE PSYCHOPHYSICAL LAWS

Various scales and relations, relating stimulus to sensation, have been recommended. As a review of the very early work of Weber and Fechner, we include the following remarks of Boring.[2] Boring's notation, for stimulus and sensation, has been altered to conform with our own.

> ... This fact properly speaking is Weber's law: if two weights differ by a just noticeable amount when separated by a given increment, then, when the weights are increased, the increment must be proportionally increased for the difference to remain just noticeable. Fechner chanced upon Weber's law and undertook to use it for the measurement of sensation. If x be the stimulus and y be the resultant sensation, and Δ signify an increment of either, then Weber's law becomes $\Delta x/x$ = a constant for the just noticeable difference. Fechner went further and assumed that all equal increments of sensation must be proportional to $\Delta x/x$, that is to say, $\Delta y = c\,\Delta x/x$ where c is a constant of proportionality. If this equation is integrated, if x be measured in terms of the threshold stimulus (the value of stimulus at which y is zero or just ready to appear), and if the constant be changed to f for common logarithms, we have $y = f \log x$.... In general it may be said that the experiments indicate that the (Weber-Fechner) law holds approximately, although not exactly, within the middle ranges of intensity, but not for very small or very large intensities.

Besides Fechner's equation, other expressions have been found relating stimulus to sensation. In particular Stevens[3] has advanced a powering relationship $y = sx^p$. (Again substituting our own notation.) The measure of x begins at threshold, s is a constant that depends on the units used and p depends on the type of stimulus, 0.3 for loudness, 0.33 to 0.5 for brightness, 1.1 for apparent length, and 4 for apparent intensity of electric shock. We do not wish to argue for any particular functional relation between stimulus and sensation but our work does assume that some such relation holds.

6.3 LINEAR MODELS

We now ask whether $\pi(x_1, x_2)$ may be used to define an additive scale on which psychological similarity can be measured. That is, does there exist a monotone function satisfying the equation

$$h\pi(x_1, x_2) + h\pi(x_2, x_3) = h\pi(x_1, x_3)?$$

[1] Fechner, G. T., *Elemente der Psychophysik*. Breitkopf and Hartel, Leipzig, 1860.
[2] Boring, E. G., Psychophysics. *Encyclopedia Britannica*, **18**, 720–723 (1955).
[3] Stevens, S. S., On the psychophysical law. *Psychological Review*, **64**, 153–181 (1957).

If so then

$$h\pi(x_1, x_2) = h\pi(x_1, x_3) - h\pi(x_2, x_3)$$
$$= t(x_1) - t(x_2)$$

where we have fixed the origin by taking x_3 as standard and we have written $t(x_1) = h\pi(x_1, x_3)$. Now

$$\pi(x_1, x_2) = H[t(x_1) - t(x_2)]$$

where H is the inverse function of h. Models of this type, with H a d.f., are called *linear* and $y_i = t(x_i)$ is called the *sensation parameter* corresponding to the stimulus x_i.

Several specific linear models have been proposed. Thurstone[4] has advanced the so-called law of comparative judgments. Sensations $\mathbf{y}_1$ and $\mathbf{y}_2$ evoked in an individual by a pair of stimuli x_1 and x_2 are assumed normal with means μ_1 and μ_2. Thurstone's model may be summarized as

$$\pi(x_1, x_2) = \int_{-(\mu_1 - \mu_2)/\sigma}^{\infty} n(t; 0, 1)\, dt \tag{6.1}$$

where σ depends on the variances and covariances of $\mathbf{y}_1$ and $\mathbf{y}_2$.

An alternative to the law of comparative judgments is the *logistic model*[5]

$$\pi(x_1, x_2) = \pi_1/(\pi_1 + \pi_2)$$
$$= L(\ln \pi_1 - \ln \pi_2) \tag{6.2}$$

where π_1 and π_2 are positive parameters without explicit significance and $L(x)$ is the *logistic* d.f.

$$L(x) = \frac{1}{1 + e^{-x}}, \quad -\infty < x < \infty.$$

Now consider a related question. If $\pi(x_1, x_2) = \frac{1}{2}$, that is, if a subject is unable to detect which is larger, then x_1 and x_2 are psychologically similar stimuli. In some cases, it may be possible to measure the psychological similarity of x_1 and x_2 by the nearness of $\pi(x_1, x_2)$ to $1/2$. It would surely prove useful if a concept of "psychological distance between stimuli" could be established having the essential properties of geometric distance between points. What models imply a distance function on the stimulus space X?

First we make precise our concept of distance. A *metric space* is a set of

[4] Thurstone, L. L., Psychophysical analysis. *Amer. J. Psychol.*, **38**, 368–389 (1927).

[5] Zermelo, E., Die Berechnung der Turnier-Ergebnisse als ein Maximumproblem der Wahrscheinlichkeitsrechnung. *Math. Zeitschrift*, **29**, 436–460 (1929). Bradley, R. A., and M. E. Terry, The rank analysis of incomplete block designs, I. The method paired comparisons. *Biometrika*, **39**, 324–345 (1952). Luce, R. Duncan, *Individual Choice Behavior*. Wiley, New York, 1959.

points X and a metric (distance function) $\rho(x_1, x_2)$ defined for all points x_1 and x_2 of X and satisfying the properties

(a) $\rho(x_1, x_2) = \rho(x_2, x_1)$
(b) $\rho(x_1, x_3) \leq \rho(x_1, x_2) + \rho(x_2, x_3)$ (triangle inequality)
(c) $\rho(x, x) = 0$
(d) $\rho(x_1, x_2) = 0$, only if $x_1 = x_2$.

If all of the above conditions with the possible exception of (d) are satisfied, the space is called *pseudometric*. If $|\pi(x_1, x_2) - \frac{1}{2}|$ is a metric or pseudometric for the stimulus space X then the corresponding paired comparison model will be called metric or pseudometric.

Theorem 6.1

A linear model is metric, if and only if,

(i) $H(-t) = 1 - H(t)$
(ii) $H(t) - H(0) \geq H(s + t) - H(s) \quad \text{for } s, t \geq 0$

and

(iii) $t(x)$ is a strictly monotone function.

Theorem 6.2

A linear model is pseudometric if and only if (i) and (ii) hold.

Relation (i) states that H is the distribution function of a symmetric random variable, (iii) is insured if larger stimuli have larger sensations, and (ii) says roughly that large sensation differences are less likely than small ones. In particular (ii) holds for absolutely continuous H with density function decreasing over the positive axis.

We prove Theorem 6.2 first. For brevity write $y_i = t(x_i)$ and note that $\rho(x_1, x_2) = |H(y_1 - y_2) - \frac{1}{2}|$. Observe that (i) implies

$$\rho(x_1, x_2) = H(|y_1 - y_2|) - \tfrac{1}{2} \tag{6.3}$$

and that, because of the monotonicity of H, (i) is equivalent to (a) and (c). Also, in the presence of (6.3), (ii) is equivalent to (b). This proves Theorem 6.2.

Preliminary to proving Theorem 6.1 we show for the linear pseudometric model that $H(t) = \frac{1}{2}$, if and only if $t = 0$. From (i) $H(0) = \frac{1}{2}$; next suppose $H(t') = \frac{1}{2}$ for $t' > 0$. From (ii) every positive interval of length t', and hence, the entire positive axis, has mass 0. Hence, $H(s) = H(0) = \frac{1}{2}$ for all $s > 0$ and H cannot be a d.f. Theorem 6.1 now follows from the equivalence of the following statements: $\rho(x_1, x_2) = 0$, $H[t(x_1) - t(x_2)] = \frac{1}{2}$, $t(x_1) = t(x_2)$. If t is strictly monotone then $\rho(x_1, x_2) = 0$ implies $x_1 = x_2$. If t is not monotone, then there exist $x_1 \neq x_2$ such that $t(x_1) = t(x_2)$ and $\rho(x_1, x_2) = 0$, contradicting (d).

Corollary 6.1

The Thurstone model (6.1) is pseudometric with sensation parameters μ_1 and μ_2. It is metric if and only if μ_i is a monotone function of stimulus x_i.
For the proof note that the function

$$H(t) = (2\pi)^{-1/2} \int_{-t/\sigma}^{\infty} \exp\left(-\frac{s^2}{2}\right) ds$$

satisfies properties (i) and (ii). In the same way, we have Corollary 6.2.

Corollary 6.2

The logistic model (6.2) is pseudometric with sensation parameters $\ln \pi_1$ and $\ln \pi_2$. It is metric, if and only if π_i is a monotone function of x_i.
The practical implication of these results is that if one succeeds in fitting data with a Thurstone, logistic, or other metric model then he can claim to have discovered a "psychological distance" for that stimulus.

7

THE CENTRAL LIMIT THEOREM

7.1 VARIANCE

The variance $V(\mathbf{x})$ of an r.v. $\mathbf{x}$ is defined in terms of the prior concept of expectation. Writing $E\mathbf{x} = \mu$ we define

$$V(\mathbf{x}) = E(\mathbf{x} - \mu)^2. \tag{7.1}$$

An immediate result, which is frequently a more convenient computing formula, is

$$V(\mathbf{x}) = E\mathbf{x}^2 - \mu^2. \tag{7.2}$$

For example, if we wish to calculate the variance of the "indicator" r.v., $\mathbf{x} = 1$ or 0 with probabilities p and q, then $\mu = 1 \cdot p + 0 \cdot q = p$; likewise, $E\mathbf{x}^2 = 1^2 p + 0^2 q = p$ and

$$V(\mathbf{x}) = p - p^2 = pq.$$

Tables 7.1 and 7.2 summarize the expectations and variances of the principal distributions which we have encountered in earlier chapters.

The practical importance of the variance is the use of its square root $D(\mathbf{x}) = V^{1/2}(\mathbf{x})$, called the *standard deviation*, to construct a probability scale. For a distribution which is nearly normal, approximately $\frac{2}{3}$, 95% and 99% of the probability measure lies between 1, 2, and 3 standard deviations, respectively. The distribution (4.22) with density

$$d(x; \alpha, \beta, \gamma) = K^{-1} \exp\left[-\frac{1}{2}\left|\frac{x - \alpha}{\gamma}\right|^{1/\beta}\right], \quad -\infty < x < \infty$$

Table 7.1 The First Two Moments of Some Discrete Distributions

Distribution	*Probability Function*	*Expectation*	*Variance*
binomial (3.8)	$\binom{n}{x} p^x(1-p)^{n-x}$; $x = 0, 1, \ldots, n$	np	$np(1-p)$
hypergeometric (3.12)	$\dfrac{\binom{D}{x}\binom{N-D}{n-x}}{\binom{N}{n}}$; $x = 0, 1, \ldots, n$	$\dfrac{nD}{N}$	$\dfrac{nD(N-D)}{N^2} \cdot \left(\dfrac{N-n}{N-1}\right)$
Poisson (3.11)	$\dfrac{\lambda^x}{x!} e^{-\lambda}$; $x = 0, 1, 2, \ldots$	λ	λ

Table 7.2 The First Two Moments of Some Absolutely Continuous Distributions

Distribution	*Probability Density*	*Expectation*	*Variance*
uniform (4.11)	$(b-a)^{-1}, a < x < b$	$\frac{1}{2}(a+b)$	$(b-a)^2/12$
normal (4.17)	$\dfrac{1}{\sqrt{2\pi}\,\sigma} \exp\left[-\dfrac{(x-\mu)^2}{2\sigma^2}\right]$	μ	σ^2
exponential (4.19)	$\rho \exp(-\rho x), \quad x \geq 0$	ρ^{-1}	ρ^{-2}
gamma (4.19)	$\dfrac{\rho(\rho x)^{\alpha-1}}{\Gamma(\alpha)} \exp(-\rho x), \quad x \geq 0$	$\dfrac{\alpha}{\rho}$	$\dfrac{\alpha}{\rho^2}$
beta (4.20)	$\dfrac{\Gamma(\mu+\nu)}{\Gamma(\mu)\cdot\Gamma(\nu)} x^{\mu-1}(1-x)^{\nu-1}$; $0 \leq x \leq 1$	$\dfrac{\mu}{\mu+\nu}$	$\dfrac{\mu\nu}{(\mu+\nu)^2(\mu+\nu+1)}$
(4.22)	$\dfrac{1}{\gamma \cdot 2^{\beta+1}\Gamma(\beta+1)} \times \exp\left[-\dfrac{1}{2}\left|\dfrac{x-\alpha}{\gamma}\right|^{1/\beta}\right]$	α	$\dfrac{\gamma^2 2^{2\beta}\Gamma(3\beta)}{\Gamma(\beta)}$

can be used to judge the accuracy of this approximation; Table 7.3 has been prepared for this purpose.

Table 7.3 $P[|\mathbf{x} - \alpha|/D(\mathbf{x}) < d]$ for the Distribution $d(x; \alpha, \beta, \gamma)$ of (4.22)

	β								
d	$\beta \to 0$ (*Uniform*)	0.2	0.4	0.5 (*Normal*)	0.6	0.8	1 (*Laplace*)	2	*Chebyshev inequality*
1	0.578	0.614	0.663	0.683	0.701	0.731	0.757	0.843	0
2	1.000	0.986	0.962	0.955	0.949	0.943	0.941	0.947	0.75
3	1.000	1.000	0.999	0.997	0.995	0.990	0.986	0.978	0.89

We next discuss a very important theoretical result known as *Chebyshev's inequality*.

Theorem 7.1

If the variance $V(\mathbf{x})$ exists then for all $K > 0$,

$$P(|\mathbf{x} - \mu| \geq K) \leq \frac{V(\mathbf{x})}{K^2}.$$

We give the proof for a continuous r.v. having density function $f(x)$. The proof for discrete r.v.'s is analogous.

$$V(\mathbf{x}) = \int_{|x-\mu|<K} (x - \mu)^2 f(x)\, dx + \int_{|x-\mu|\geq K} (x - \mu)^2 f(x)\, dx$$

$$\geq \int_{|x-\mu|\geq K} (x - \mu)^2 f(x)\, dx \geq K^2 \int_{|x-\mu|\geq K} f(x)\, dx = K^2 P(|\mathbf{x} - \mu| \geq K).$$

Corollary 7.1

If $V(\mathbf{x}) = 0$ then $P(\mathbf{x} = \mu) = 1$.

Proof

For every $\epsilon > 0$, $P(|x - \mu| < \epsilon) = 1 - P(|\mathbf{x} - \mu| \geq \epsilon) \geq 1 - V(\mathbf{x})/K^2$. Therefore, $P(|\mathbf{x} - \mu| < \epsilon) = 1$, $\epsilon > 0$. The corollary follows from the continuity property on taking the limit as $\epsilon \to 0$.

Chebyshev's inequality cannot be improved using solely the first two moments; for every K there is a distribution, having given moments μ and σ^2, which achieves equality in Chebyshev's result. Specifically, equality holds for the distribution

$$P(\mathbf{x} = \mu \pm K) = \frac{\sigma^2}{2K^2} \quad \text{and} \quad P(\mathbf{x} = \mu) = 1 - \frac{\sigma^2}{K^2}.$$

But comparison of Chebyshev's inequality with the normal distribution (Table 7.3) shows that the inequality is frequently a liberal bound. The result is primarily of theoretical importance. Also, it gives meaning to the concept of variance as a measure of dispersion for arbitrary distributions; when the variance is small, the probability of large deviations from expectation must necessarily be small.

The Russian probabilists consider Chebyshev to be the founder of their modern school. They speak of "the Chebyshev tradition" which they characterize by the introduction of rigor, structure, and the methods of analysis into probability. They believe that the Chebyshev tradition has produced, out of a class of individual problems, a basic probability theoretic scheme capable of handling a large number of concrete problems.

In Theorem 3.2 we proved that E is a linear operator; V is not linear but equally useful formulas may be developed.

Theorem 7.2

If a and b are constants while $\mathbf{x}$ and $\mathbf{y}$ are r.v.'s with $E\mathbf{x} = \mu_x$ and $E\mathbf{y} = \mu_y$ then

$$V(a\mathbf{x} + b) = a^2 V(\mathbf{x}) \tag{7.3}$$

$$V(\mathbf{x} + \mathbf{y}) = V(\mathbf{x}) + V(\mathbf{y}) + 2E(\mathbf{x} - \mu_x)(\mathbf{y} - \mu_y). \tag{7.4}$$

We perform the necessary computations only for Equation (7.4).

$$\begin{aligned} V(\mathbf{x} + \mathbf{y}) &= E[\mathbf{x} + \mathbf{y} - (\mu_x + \mu_y)]^2 = E[(\mathbf{x} - \mu_x) + (\mathbf{y} - \mu_y)]^2 \\ &= E(\mathbf{x} - \mu_x)^2 + E(\mathbf{y} - \mu_y)^2 + 2E(\mathbf{x} - \mu_x)(\mathbf{y} - \mu_y) \\ &= V(\mathbf{x}) + V(\mathbf{y}) + 2E(\mathbf{x} - \mu_x)(\mathbf{y} - \mu_y). \end{aligned}$$

In analogy with the variance formula, the quantity $E(\mathbf{x} - \mu_x)(\mathbf{y} - \mu_y)$ is called the *covariance* $C(\mathbf{x}, \mathbf{y})$ of $\mathbf{x}$ and $\mathbf{y}$. Corresponding to the results of Theorem 7.2 we have

$$C(a_1\mathbf{x}_1 + a_2\mathbf{x}_2, b_1\mathbf{x}_1 + b_2\mathbf{x}_2) = a_1b_1V(\mathbf{x}_1) + a_2b_2V(\mathbf{x}_2) + (a_1b_2 + a_2b_1)C(\mathbf{x}_1, \mathbf{x}_2). \tag{7.5}$$

The much discussed *correlation coefficient* is a normalized covariance; the correlation between $\mathbf{x}$ and $\mathbf{y}$ is given by $R(\mathbf{x}, \mathbf{y}) = C(\mathbf{x}, \mathbf{y})/\sqrt{V(\mathbf{x}) \cdot V(\mathbf{y})}$. Two variables are *uncorrelated* if their correlation coefficient is 0 or equivalently if their covariance is 0. Equation (7.4) yields Corollary 7.2.

Corollary 7.2

If **x** and **y** are uncorrelated then $V(\mathbf{x} + \mathbf{y}) = V(\mathbf{x}) + V(\mathbf{y})$.

Theorem 7.3

$C^2(\mathbf{x}, \mathbf{y}) \leq V(\mathbf{x}) \cdot V(\mathbf{y})$ or equivalently (if $V(\mathbf{x}) \cdot V(\mathbf{y}) \neq 0$) $|R(\mathbf{x}, \mathbf{y})| \leq 1$, equality holding only if **x** and **y** lie on a straight line with probability 1. The slope of this line is $+$ or $-$ according as $R = +1$ or -1.

Proof

$V(a\mathbf{x} + b\mathbf{y}) = a^2V(\mathbf{x}) + b^2V(\mathbf{y}) + 2abC(\mathbf{x}, \mathbf{y}) \geq 0$ for arbitrary a and b. Taking $a = V(\mathbf{x})^{-1/2}$ and $b = \pm V(\mathbf{y})^{-1/2}$, we obtain $2[1 \pm R(\mathbf{x}, \mathbf{y})] \geq 0$. If $R = \pm 1$, then $V[\mathbf{x}V(\mathbf{x})^{-1/2} \mp \mathbf{y}V(\mathbf{y})^{-1/2}] = 0$ and, from Corollary 7.1, $\mathbf{y} = \pm [V(\mathbf{y})/V(\mathbf{x})]^{1/2}\mathbf{x}$ plus a constant, with probability 1.

Theorem 7.4

If **x** and **y** are independent r.v.'s whose variances exist then they are uncorrelated.

Proof

We perform the needed computations only for continuous r.v.'s.

$$C(\mathbf{x}, \mathbf{y}) = \int_{-\infty}^{\infty} \int_{-\infty}^{\infty} (x - \mu_x)(y - \mu_y)g(x) \cdot h(y)\, dx\, dy$$

where $g(x)$ and $h(y)$ are the densities of **x** and **y**.

$$C(\mathbf{x}, \mathbf{y}) = \int_{-\infty}^{\infty} (x - \mu_x)g(x)\, dx \cdot \int_{-\infty}^{\infty} (y - \mu_y)h(y)\, dy$$

each of these integrals is 0, and hence, their product is 0.

The converse result is not true as may be seen by considering the sum and the difference of the dots on a red and green die. From Table 2.1, we see that these two random variables are dependent; however, their correlation is 0.

A continuous counterexample to the converse of Theorem 7.4 is the following. A unit of mass is uniformly distributed on a disk (circle with interior) centered at the origin in the (x, y) plane. Clearly no two marginal distributions of **x** and **y** will multiply together to give this joint distribution. Hence, **x** and **y** are dependent. However, because of symmetry, $E\mathbf{xy} = E\mathbf{x} = E\mathbf{y} = 0$, showing that **x** and **y** are uncorrelated.

Corollary 7.3

If $\mathbf{x}$ and $\mathbf{y}$ are independent then $V(\mathbf{x} + \mathbf{y}) = V(\mathbf{x}) + V(\mathbf{y})$. This follows from Theorem 7.4 and Corollary 7.2.

The mean and variance are by far the most commonly used moments but others can be defined and sometimes are of value. We define:

the kth *moment*

$$\xi_k = E\mathbf{x}^k, \quad k = 1, 2, \ldots;$$

the kth *central moment*

$$\mu_k = E(\mathbf{x} - \mu)^k, \quad k = 1, 2, \ldots;$$

and the kth *absolute moment*

$$\nu_k = E|\mathbf{x}|^k, \quad k = 1, 2, \ldots.$$

Clearly $\xi_1 = E\mathbf{x} = \mu$, $\mu_2 = V(\mathbf{x})$, and $\nu_k = \xi_k$ for k even. The moments and central moments can be expressed in terms of one another; we have $\mu_2 = \xi_2 - \xi_1^2$, $\mu_3 = \xi_3 - 3\xi_1\xi_2 + 2\xi_1^3$, and so forth.

7.2 SUMS OF INDEPENDENT AND IDENTICALLY DISTRIBUTED R.V.'s

Theorem 7.5

If

$$\bar{\mathbf{x}} = \frac{1}{n}(\mathbf{x}_1 + \cdots + \mathbf{x}_n),$$

where the $\{\mathbf{x}_i\}$ are independent and identically distributed with common mean μ and common variance σ^2, then

$$P(|\bar{\mathbf{x}} - \mu| > \epsilon) \to 0.$$

Proof

$$E\bar{\mathbf{x}} = \mu. \quad V(\bar{\mathbf{x}}) = \sigma^2/n. \quad P(|\bar{\mathbf{x}} - \mu| > \epsilon) \leq \sigma^2/n\epsilon^2.$$

The above theorem is just one of many "Laws of Large Numbers." In application these laws are interpreted as saying that the average of a large number of independent identically distributed r.v.'s is likely to differ but little from the common mean.

The special case of Theorem 7.5 where one is treating a sequence of Bernoulli trials and $\mathbf{x}_i$ is the indicator for the ith trial, that is, $\mathbf{x}_i = 1$ or 0 according as S or F occurs on the ith trial, is called Bernoulli's theorem. We have $\mu = p$ and $\sigma^2 = pq$. Thus, Bernoulli's theorem says that $\bar{\mathbf{x}}$, the proportion of observed S's, approaches p, the probability of S on a single trial, in probability. Bernoulli's theorem was the forerunner of all laws of large numbers. The version given in Theorem 7.5 is, of course, due to Chebyshev.

Next we consider two independent r.v.'s $\mathbf{x}$ and $\mathbf{y}$ which have d.f.'s $F(x)$ and $G(y)$. Required, to find the distribution $H(x)$ of the r.v. $\mathbf{z} = \mathbf{x} + \mathbf{y}$. The density functions of all r.v.'s are assumed to exist and are denoted by corresponding lower case letters. From the theorem on total probability,

$$P(\mathbf{z} \leq z, a < \mathbf{x} \leq b) = \sum_{i=1}^{k} P(\mathbf{x} + \mathbf{y} \leq z | x_i < \mathbf{x} \leq x_{i+1}) \cdot P(x_i < \mathbf{x} \leq x_{i+1})$$

for all partitions $a = x_1 < x_2 < \cdots < x_{k+1} = b$ of the interval (a, b). Allowing k to approach ∞ in such a way that $\max (x_{i+1} - x_i)$ goes to 0, we obtain

$$P(\mathbf{z} \leq z, a < \mathbf{x} \leq b) = \int_a^b G(z - x) f(x)\, dx$$

$$H(z) = P(\mathbf{z} \leq z) = \int_{-\infty}^{\infty} G(z - x) f(x)\, dx \tag{7.6}$$

and

$$h(z) = H'(z) = \int_{-\infty}^{\infty} g(z - x) f(x)\, dx \tag{7.7}$$

where strictly speaking the differentiation under the integral sign needs some justification; which we omit. The *convolution formulas* (7.6) and (7.7) are fundamental and omnipresent in the applications.

For example, let $\mathbf{x}_1, \mathbf{x}_2, \ldots$ be a sequence of computational roundoff errors; in the spirit of Section 4.3 we consider them as independent r.v.'s with common uniform density function

$$f(x) = \begin{cases} 1, & -\frac{1}{2} \leq x \leq \frac{1}{2} \\ 0, & \text{otherwise} \end{cases}$$

and let $f^{(n)}(x)$ be the density of $\mathbf{x}_1 + \mathbf{x}_2 + \cdots + \mathbf{x}_n$. We may show that

$$f^{(n+1)}(x) = \int_{x-1/2}^{x+1/2} f^{(n)}(t)\, dt$$

and in particular

$$f^{(2)}(x) = \begin{cases} 1 + x, & -1 \le x \le 0 \\ 1 - x, & 0 \le x \le 1 \\ 0, & \text{otherwise} \end{cases} \tag{7.8}$$

$$f^{(3)}(x) = \begin{cases} \frac{1}{2}(\frac{3}{2} + x)^2, & -\frac{3}{2} \le x \le -\frac{1}{2} \\ \frac{3}{4} - x^2, & -\frac{1}{2} \le x \le \frac{1}{2} \\ \frac{1}{2}(\frac{3}{2} - x)^2, & \frac{1}{2} \le x \le \frac{3}{2} \\ 0, & \text{otherwise.} \end{cases}$$

The mean and variance of the distribution with density $f^{(n)}(x)$ are 0 and $n/12$, respectively. Figure 7.1 is a comparison of $f^{(n)}(x)$ with a normal having the same variance. In the next section, we will treat the central limit theorem which implies that $f^{(n)}(x)$ tends towards normality as n becomes infinite (in the sense that areas under the two curves become equal). From Figure 7.1, we see that the approximation is already pretty good for $n = 3$.

7.3 OPERATIONAL METHODS IN PROBABILITY

Operational methods were first applied to probability problems by James Bernoulli who observed that the number of ways of obtaining m in throwing n dice is the coefficient of x^m in the expansion of $(x + x^2 + x^3 + x^4 + x^5 + x^6)^n$. Thus, Bernoulli solved a probability problem by transforming it into a related one concerning a function which at first seems to have nothing to do with the probability problem. Such techniques are called operational or transform methods; they are used in many branches of mathematics, particularly in solving differential equations. The kind of transform introduced by Bernoulli is now called a *generating function*.

The method of characteristic functions is an indirect operational technic which is immensely important in modern probability theory. Later, we will use this method to prove the central limit theorem.

The *characteristic function* (*c.f.*) of an r.v. $\mathbf{x}$ is defined as

$$\phi_{\mathbf{x}}(t) = E \cos t\mathbf{x} + iE \sin t\mathbf{x}$$

or more briefly, extending the definition of expectation to complex variables and dropping the subscript $\mathbf{x}$, as

$$\phi(t) = E \exp(it\mathbf{x}). \tag{7.9}$$

In particular if $\mathbf{x}$ has density function $f(x)$ then (7.9) becomes

$$\phi(t) = \int_{-\infty}^{\infty} \exp(itx) f(x)\, dx.$$

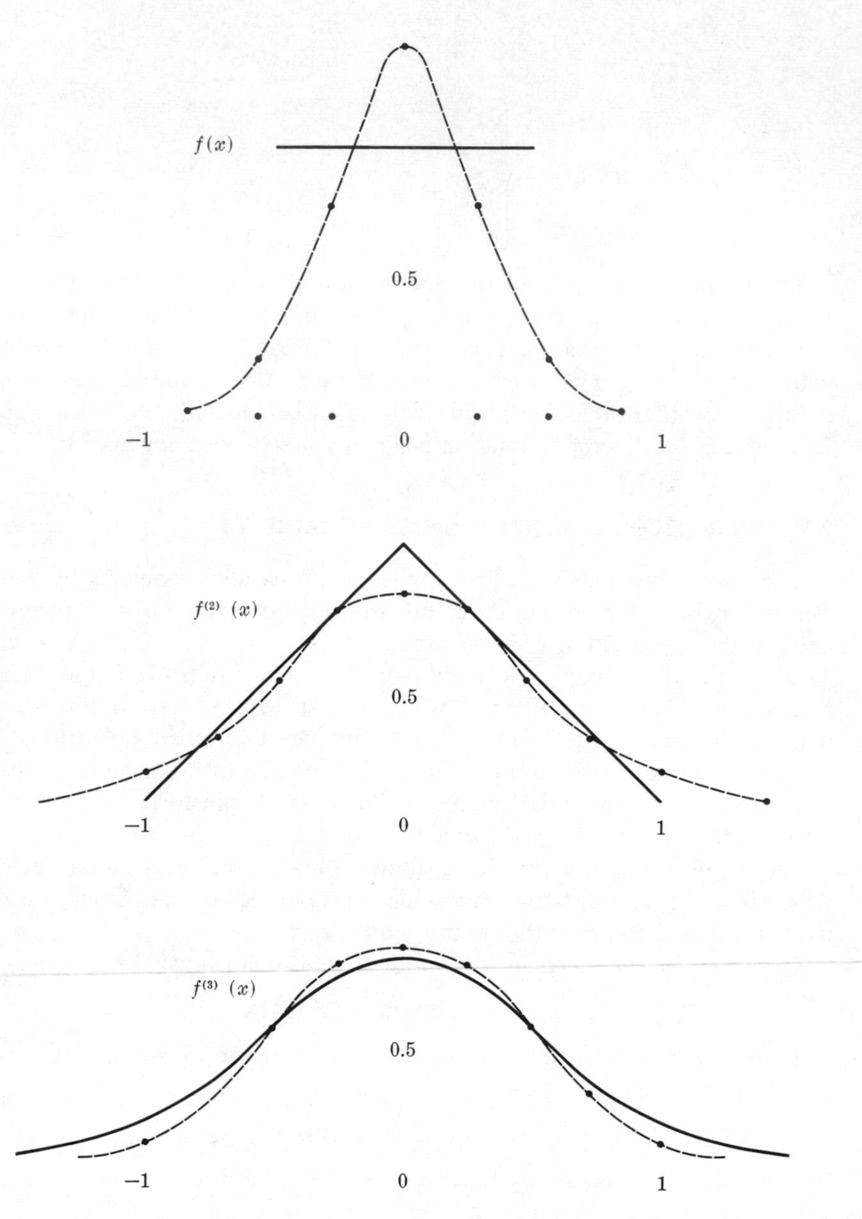

Figure 7.1 Comparison of $f^{(n)}(x)$ with a Normal Density (Dotted) Having the Same Variance

This relationship is expressed mathematically by saying that $\phi(t)$ is the Fourier transform of $f(x)$.

Moments can be obtained by formal differentiation of (7.9).[1] If $E\mathbf{x}^k$ exists then

$$\phi^{(j)}(0) = i^j \int x^j f(x)\,dx = i^j E\mathbf{x}^j, \quad j = 0, 1, \ldots k$$

and a formal MacLaurin expansion is possible:

$$\phi(t) = \sum_{j=0}^{k} \frac{\phi^{(j)}(0)}{j!} t^j + o(t^k) = 1 + \sum_{j=1}^{k} \frac{(it)^j}{j!} E\mathbf{x}^j + o(t^k).$$

Much of the usefulness of c.f.'s is due to the following two results concerning linear combinations of r.v.'s.

Theorem 7.6

If $\mathbf{y} = a\mathbf{x} + b$, then

$$\phi_{\mathbf{y}}(t) = e^{itb} \cdot \phi_{\mathbf{x}}(at).$$

Proof

$$\phi_{\mathbf{y}}(t) = Ee^{it(a\mathbf{x}+b)} = e^{itb}Ee^{i(at)\mathbf{x}} = e^{itb}\phi_{\mathbf{x}}(at).$$

Theorem 7.7

The c.f. of the sum of two independent r.v.'s is the product of their c.f.'s.

Proof

$$Ee^{it(\mathbf{x}+\mathbf{y})} = Ee^{it\mathbf{x}} \cdot Ee^{it\mathbf{y}}.$$

Two advanced results concerning characteristic functions will be needed. We may summarize these loosely as follows. There is a one-one correspondence between distribution and characteristic functions which is preserved under limit operations. More precise statements follow.

Theorem 7.8

A distribution function is uniquely determined by its characteristic function.

[1] Cramér, Harald, *Mathematical Methods of Statistics*. Princeton University Press, Princeton, N.J., 1946.

The proof consists of showing that if $\phi(t)$ and $F(x)$ are c.f. and d.f. of $\mathbf{x}$, and if $a - h$ and $a + h$ are points of continuity of $F(x)$, then

$$F(a + h) - F(a - h) = \frac{1}{\pi} \lim_{c \to \infty} \int_{-c}^{c} \frac{\sin ht}{t} e^{-ita} \phi(t)\, dt.$$

We prove this formula for discrete r.v.'s; the proof in the general case is analogous but requires more machinery. If $\mathbf{x}$ is discrete, then

$$\phi(t) = \sum_{X} e^{itx} p(x)$$

and

$$\begin{aligned} K = \int_{-c}^{c} \frac{\sin ht}{t} e^{-ita} \phi(t)\, dt &= \sum_{X} p(x) \int_{-c}^{c} \frac{\sin ht}{t} e^{it(x-a)}\, dt \\ &= \sum_{X} p(x) \cdot 2 \int_{0}^{c} \frac{\sin ht}{t} \cdot \cos t(x - a)\, dt \\ &= \sum_{X} p(x) \int_{0}^{c} \frac{\sin t(x - a + h) - \sin t(x - a - h)}{t}\, dt. \end{aligned}$$

Using the calculus formula

$$\lim_{c \to \infty} \int_{0}^{c} \frac{\sin t}{t} = \frac{\pi}{2}$$

we obtain

$$\lim_{c \to \infty} \int_{0}^{c} \frac{\sin t(x - a + h) - \sin t(x - a - h)}{t}\, dt = \begin{cases} 0, & x < a - h \\ \pi/2, & x = a - h \\ \pi, & a - h < x < a + h \\ \pi/2, & x = a + h \\ 0, & x > a + h. \end{cases}$$

Also, since $F(x)$ is continuous at $a - h$ and $a + h$,

$$\lim_{c \to \infty} K = \pi \sum_{a-h<x<a+h} p(x) = \pi[F(a + h) - F(a - h)]$$

which yields the above inversion formula.

The second advanced result that we will need is a continuity theorem which we state without proof.

Theorem 7.9

Given d.f.'s $F(x)$, $F_1(x)$, $F_2(x), \ldots$ with corresponding c.f.'s $\phi(t)$, $\phi_1(t)$, $\phi_2(t), \ldots$, then $F_n(x) \to F(x)$ at points of continuity of $F(x)$ if and only if $\phi_n(t) \to \phi(t)$ for every t.

We shall see that if $\mathbf{y}_n = \mathbf{x}_1 + \cdots + \mathbf{x}_n$ then under very general conditions

$$P\left(\frac{\mathbf{y}_n - E\mathbf{y}_n}{D(\mathbf{y}_n)} \leq y\right) \xrightarrow[n\to\infty]{} (2\pi)^{-1/2} \int_{-\infty}^{y} e^{-x^2/2}\,dx = N(y). \qquad (7.10)$$

Various results, giving conditions under which the above limit holds, are called *central limit theorems.*

To prove a central limit theorem we need to know explicitly the c.f. of the normal distribution.

Theorem 7.10

The c.f. of an r.v. having normal density $n(x; 0, 1)$ is $\exp(-t^2/2)$.

Proof

Write

$$z(t) = \int \cos tx e^{-x^2/2}\,dx + i\int \sin tx e^{-x^2/2}\,dx$$

$$z'(t) = -\int x \sin tx e^{-x^2/2}\,dx + i\int x \cos tx e^{-x^2/2}\,dx.$$

The imaginary part of $z'(t)$ is 0, since its integrand is an odd function. Hence, the imaginary part of $z(t)$ is constant, but $z(0) = \sqrt{2\pi}$ so that $z(t)$ is a real function and

$$z'(t) = -\int x \sin tx e^{-x^2/2}\,dx$$

$$= -t\int \cos tx e^{-x^2/2}\,dx = -t\cdot z(t)$$

after an integration by parts. Solving this differential equation and using the relation $z(0) = \sqrt{2\pi}$ to evaluate the constant, we obtain the theorem.

We can now prove the following simple and yet quite general central limit theorem.

Theorem 7.11

If the independent r.v.'s $\mathbf{x}_1, \mathbf{x}_2, \ldots$ are identically distributed and have nonzero finite variances, then the limit (7.10) holds for all y.

Proof

Define μ and σ^2 to be the mean and variance of $\mathbf{x}_k$. Let $\phi(t)$ be the c.f. of $\mathbf{x}_k - \mu$ and write $\phi(t) = 1 - \frac{1}{2}\sigma^2 t^2 + o(t^2)$. The c.f. of $(\mathbf{y}_n - E\mathbf{y}_n)/D(\mathbf{y}_n)$ is

$$\left[1 - \frac{t^2}{2n} + o\left(\frac{1}{n}\right)\right]^n$$

which approaches $\exp(-t^2/2)$, the c.f. of the normal distribution. Because of the unique and continuous correspondence between characteristic and distribution functions, the theorem is proved.

A more careful statement of the above approach to $\exp(-t^2/2)$ is contained in Lemma 7.1.

Lemma 7.1

$$\lim_{n\to\infty} [1 + b/n + o(1/n)]^n = e^b, \text{ for } b \text{ real.}$$

Proof

Recall that we write $y(x) = o(x)$, if $y(x)/x \to 0$, as $x \to 0$. Taylor's expansion with remainder can be found in the following form in most calculus books.

$$\begin{aligned} y(x) &= y(0) + y'(\theta x)x \\ &= y(0) + y'(0)x + [y'(\theta x) - y'(0)]\, x \end{aligned}$$

where $0 < \theta < 1$. Hence, if $y'(x)$ is continuous at the origin, then

$$y(x) = y(0) + y'(0)x + o(x).$$

Now take $y(x) = \log(1 + x)$ and note that $y'(x) = (1 + x)^{-1}$ is continuous at $x = 0$. We find

$$\log(1 + x) = x + o(x).$$

$$\begin{aligned} n \log\left[1 + \frac{b}{n} + o\left(\frac{1}{n}\right)\right] &= n\left\{\frac{b}{n} + o\left(\frac{1}{n}\right) + o\left[\frac{b}{n} + o\left(\frac{1}{n}\right)\right]\right\} \\ &= b + no\left(\frac{1}{n}\right) = b + o(1). \end{aligned}$$

Raising both sides to a power we get

$$\left[1 + \frac{b}{n} + o\left(\frac{1}{n}\right)\right]^n = e^{b + o(1)}$$

which proves the lemma.

The first central limit theorem was proved by Abraham De Moivre around 1733 and was first published in his *Doctrine of Chances*. De Moivre used Stirling's approximation to treat Bernoulli trials for the case $p = \frac{1}{2}$. Letting $\mathbf{x}_i$ be the indicator of the ith trial then $\mathbf{y}_n$ is the number of S's in n trials. De Moivre's theorem is then: For $p = \frac{1}{2}$

$$P\left(\frac{\mathbf{y}_n - n/2}{(n/4)^{1/2}} \leq y\right) \to N(y).$$

Laplace subsequently extended De Moivre's theorem to the case of arbitrary p.

De Moivre was born in France, but went to England in 1685 to escape religious persecution. There he earned his living by teaching mathematics and serving as a consultant on actuarial and other problems of a probabilistic nature. In 1865, Todhunter judged that "the theory of probability owes more to (De Moivre) than to any other mathematician, with the sole exception of Laplace."

Marquis Pierre-Simon de Laplace (1749–1827) was first a mathematical astronomer. His work in probability, contained in the *Théorie Analytique des Probabilités*, was inspired by a need for it in astronomy. In addition to his work on the central limit theorem, Laplace set the entire subject of probability forward as an organized theory. He treated least squares, Bayes' Principle, statistics, and provided elegant solutions to all of the important results known in his day.

Several transforms besides the characteristic function are useful in probability. Two of these are the *moment generating function* defined by

$$m(t) = E \exp t\mathbf{x},$$

provided that the defining integral exists in some interval $-\epsilon < t < \epsilon$, and the related *Laplace transform* used only for positive r.v.'s and defined as

$$L(t) = E \exp(-t\mathbf{x}).$$

The main advantage of the c.f. over the moment generating function is a theoretical one. The c.f. is always well defined since sin and cos are bounded functions but the integral defining $m(t)$ may diverge. Thus, if one is working rigorously, he must constantly be checking the existence of $m(t)$, a question which does not arise for $\phi(t)$. The Laplace transform has the same advantage when one is working with nonnegative r.v.'s since $L(t)$ will converge for all $t \geq 0$.

We will find Laplace transforms useful in the last chapter in connection with renewal theory.

PROBLEMS

1. Prove that $P(|\mathbf{x} - \mu| > \epsilon) \leq \mu_4/\epsilon^4$, for $\epsilon > 0$ where $\mathbf{x}$ is an r.v. with finite fourth central moment μ_4 and with mean μ.

2. Show that the area under the Cauchy density function, $c(x) = [\pi(1 + x^2)]^{-1}$, is 1.

3. Show that if $\boldsymbol{\alpha}$ is uniformly distributed in the interval 0 to 2π that $\mathbf{y} = \tan \boldsymbol{\alpha}$ has the Cauchy distribution.

4. If male and female births are equally likely then what is the probability of 106 or more boys among 200 newborn children?

5. Calculate the mean and variance of an r.v. having the density

$$f(x) = \begin{cases} |x|, & |x| \leq 1 \\ 0, & |x| > 1. \end{cases}$$

6. A sequence of distribution functions $\{F_n(x)\}$ can converge to a d.f. $F(x)$ without the corresponding sequence of variances $V(\mathbf{x}_n)$ converging to the variance of $F(x)$. Show this using the sequence of probability functions $P(\mathbf{x}_n = 0) = \frac{1}{2}$, $P(\mathbf{x}_n = 1) = \frac{1}{2} - 2^{-2n}$, and $P(\mathbf{x}_n = 2^n) = 2^{-2n}$ for $n = 1, 2, \ldots$.

7. Let $\mathbf{x}$ be an r.v. with $P[\mathbf{x} = 1] = \frac{3}{4}$, $P[\mathbf{x} = \frac{1}{5}] = \frac{1}{4}$. Let the random variable $\mathbf{z}$ be defined by the equation $\mathbf{z} = 1/\mathbf{x}$.

   (a) Find $E\mathbf{z}$.
   (b) Find $V(\mathbf{z})$.
   (c) Find the joint distribution of $\mathbf{x}$ and $\mathbf{z}$.
   (d) Find the covariance of $\mathbf{x}$ and $\mathbf{z}$.

8. Let $\mathbf{x}$ be an r.v. with the geometric distribution. That is, assume $P[\mathbf{x} = k] = pq^k$, $k = 0, 1, \ldots$.

   (a) Find the probability generating function of $\mathbf{x}$.
   (b) Find $E\mathbf{x}$.
   (c) Find the generating function for $\mathbf{x} + 2$.

9. Let $\mathbf{x}_i$ be a sequence of independent random variables, with $E\mathbf{x}_i = 0$. Suppose that there is a random variable $\mathbf{y}$ such that $E\mathbf{y} = 0$, $E\mathbf{y}^2 < \infty$ and $|\mathbf{x}_i| < \mathbf{y}$ (with probability one). Show that the law of large numbers holds for the sequence $\{\mathbf{x}_i\}$. (*Hint*: Use the Chebyshev Inequality.)

10. Recall that Martian coins are three sided (heads, tails, and torsos). Assume a fair coin (each side has a probability of $\frac{1}{3}$ of appearing on any toss) and independent tosses.

    (a) What is the probability that in 10 tosses exactly 3 heads appear?
    (b) What is the probability that in 3 tosses, one head, one tail, and one torso appear?
    (c) Approximate the probability that in 1800 tosses the number of heads appearing is 600 or less.

11. Consider the proposition below. If it is true, prove it is true. If it is false, give a counterexample to show that it is false. Then add more hypotheses to the proposition to make it true, and prove that it is in fact true. The proposition is:

    Let b be a positive real number and $\mathbf{y}$ a random variable. Then

$$P[\mathbf{y} > b] \leq E\mathbf{y}/b.$$

12. Let **x** and **y** be random variables with joint probability function $P(x_j, y_k) = P[\mathbf{x} = x_j, \mathbf{y} = y_k]$. Suppose $P(1, 1) = \frac{1}{8}$, $P(1, -1) = \frac{5}{16}$, $P(0, 1) = \frac{3}{8}$, $P(-1, -1) = \frac{3}{16}$.

    (a) Find the probability function for $\mathbf{z} = \mathbf{x} + \mathbf{y}$.
    (b) Find $E\mathbf{z}$.
    (c) Find the covariance of **x** and **y**.
    (d) Are **x** and **y** independent? Support your answer with the statement of a pertinent theorem or definition.

13. The random variable **x** has a density function

$$f(x) = \begin{cases} 3x(2 - x)/4, & \text{for } 0 < x < 2 \\ 0, & \text{otherwise.} \end{cases}$$

Show that the distribution is symmetrical, with mean 1 and variance $\frac{1}{5}$. Show that the second and third moments about $x = 0$ are $\frac{6}{5}$ and $\frac{8}{5}$, respectively.

14. Prove that a characteristic function is uniformly continuous over the whole real line and satisfies the relations:

$$|\phi(t)| \le \phi(0) = 1.$$

8

*A UNIFIED APPROACH TO PROBABILITY

8.1 EXPECTATION AS A LEBESGUE INTEGRAL

Separate definitions of expectation have previously been given for discrete and continuous r.v.'s. A general theory containing these two definitions as special cases will now be given. This is a theoretical topic which may be omitted by a reader who is interested only in applications. However, the new concept of expectation proves very useful from a theoretical point of view and much of the literature of probability is written in this general context.

We first briefly discuss general measures. Our main concern is probability but there are other important measures; volume is an example. Given a universal set S of objects denoted individually by s and a class $\mathscr{A}$ of subsets of S which is closed under countable set operations then one may define a measure for the sets of $\mathscr{A}$. A set function $\mu(A)$ is called a measure if

(i) $\mu(A) \geq 0$ for all $A \in \mathscr{A}$.

(ii) $\mu(A_1 + A_2 + \cdots) = \mu(A_1) + \mu(A_2) + \cdots$ for all countable collections $\{A_i\}$ of pairwise disjoint sets of $\mathscr{A}$.

(iii) $\mu(F)$ is finite for some $F \in \mathscr{A}$.

If, as in the case of probability, $\mu(S)$ is finite then $\mu(A)$ is a *finite measure*. For any measure, $\mu(\Phi) = 0$ since $F + \Phi = F$ and $\mu(F) + \mu(\Phi) = \mu(F)$. We sometimes speak of the measure $(S, \mathscr{A}, \mu)$ since all three of these things are essential.

We now define the Lebesgue integral for real-valued measurable functions. A real-valued function $\mathbf{x}(s)$ is *measurable* if $\{s: \mathbf{x}(s) \leq x\} \in \mathscr{A}$, for all real x. In defining the Riemann integral of ordinary calculus one subdivides the domain of the function to be integrated, forms sums based on a step function approximation, and takes the limit of these sums as the division becomes finer. For the Lebesgue integral the process is similar, the essential difference being that one subdivides the range instead of the domain.

For an arbitrary measurable function $\mathbf{x}(s)$, we form the sums

$$S_n = \sum_{k=1}^{n2^n} \frac{k-1}{2^n}\,\mu\left[\frac{k-1}{2^n} < \mathbf{x}(s) \leq \frac{k}{2^n}\right] + n\cdot\mu[\mathbf{x}(s) > n]$$

$$T_m = \sum_{k=1}^{m2^m} \frac{k}{2^m}\,\mu\left[-\frac{k}{2^m} < \mathbf{x}(s) \leq -\frac{k-1}{2^m}\right] + m\cdot\mu[\mathbf{x}(s) \leq -m].$$

The *Lebesgue integral* of $\mathbf{x}$ with respect to the measure μ is then defined to be

$$\int \mathbf{x}\, d\mu = \lim_{n\to\infty} S_n - \lim_{m\to\infty} T_m$$

provided both terms in the defining difference are finite. If both terms of the difference are infinite, we say that the integral does not exist; but if only one term is infinite we say that the integral diverges to $+\infty$ or $-\infty$ according as the first or second term is the infinite one.

We may describe the integral which we have discussed thus far as the definite integral. An indefinite integral of $\mathbf{x}$ over $A \subset S$ is defined as

$$\int_A \mathbf{x}\, d\mu = \int \mathbf{x}\mathbf{i}_A(s)\, d\mu$$

where

$$\mathbf{i}_A(s) = \begin{cases} 1, & s \in A \\ 0, & s \notin A. \end{cases}$$

In particular,

$$\int_A d\mu = \mu(A).$$

The abstract definition of an r.v. $\mathbf{x}$ is that it is a measurable function with respect to a probability measure. The measurability condition is needed so that the d.f. of $\mathbf{x}$ will always be defined. A general concept of expectation, embracing both discrete and continuous r.v.'s, is now defined in terms of the Lebesgue integral:

$$E\mathbf{x} = \int \mathbf{x}\, dP. \quad \text{In particular, } E\mathbf{i}_A = P(A).$$

An alternative notation, emphasizing the distribution function rather than the probability measure, is

$$E\mathbf{x} = \int x\, dF(x)$$

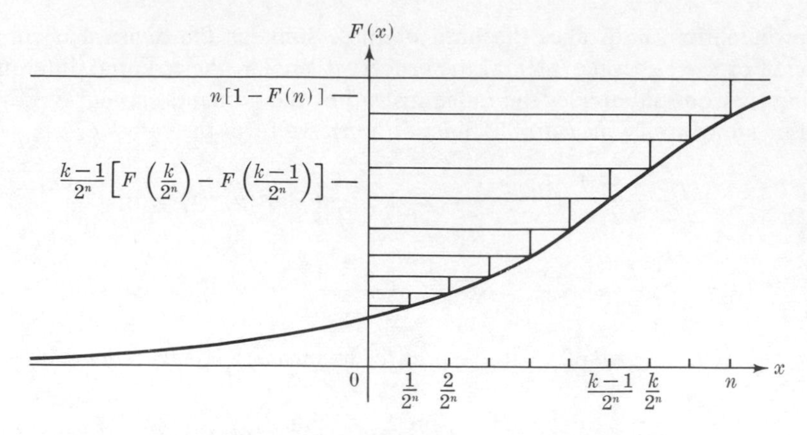

Figure 8.1 Areas Contributing to the Sum S_n

where $F(x)$ is the distribution function of **x**. The appropriateness of this notation is clear from the equations

$$S_n = \sum_{k=1}^{n2^n} \frac{k-1}{2^n}\left[F\left(\frac{k}{2^n}\right) - F\left(\frac{k-1}{2^n}\right)\right] + n[1 - F(n)]$$

and

$$T_m = \sum_{k=1}^{m2^m} \frac{k}{2^m}\left[F\left(-\frac{k-1}{2^m}\right) - F\left(-\frac{k}{2^m}\right)\right] + m \cdot F(-m).$$

A graphic interpretation is useful here. S_n is the sum of the areas of the rectangles of Figure 8.1. Thus, if S_n approaches a finite limit S, then S is the

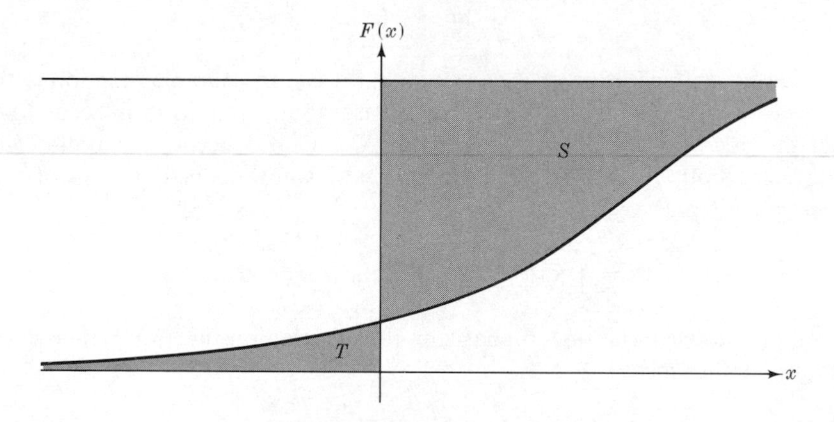

Figure 8.2 $E\mathbf{x} = S - T$

area indicated in Figure 8.2. Similarly lim T_m is the area marked T. Several conclusions are clear from this graphic interpretation.

First, the graph yields an alternative formula for evaluating $E\mathbf{x}$. We have

$$E\mathbf{x} = S - T = \int_0^\infty [1 - F(x)]\,dx - \int_{-\infty}^0 F(x)\,dx. \tag{8.1}$$

This formula is sometimes the easiest to apply.

Second, the graph makes it obvious that many methods of partition, besides the one we chose, will yield the expectation. If $E\mathbf{x}$ exists then

$$\lim_{n\to\infty} \sum_{k=1}^{n2^n} \frac{k-1}{2^n}\left[F\left(\frac{k}{2^n}\right) - F\left(\frac{k-1}{2^n}\right)\right] = S$$

and hence, $n[1 - F(n)]$ approaches 0. Similarly $\lim_{n\to\infty} mF(-m) = 0$. Further methods of partition also yield the same limits. For example,

$$S = \lim_{n\to\infty} \sum_{k=1}^{n2^n} \xi_k\left[F\left(\frac{k}{2^n}\right) - F\left(\frac{k-1}{2^n}\right)\right], \quad \frac{k-1}{2^n} \leq \xi_k \leq \frac{k}{2^n}$$

$$T = \lim_{m\to\infty} \sum_{k=1}^{m2^m} \xi_k\left[F\left(-\frac{k-1}{2^m}\right) - F\left(-\frac{k}{2^m}\right)\right], \quad \frac{k-1}{2^m} \leq \xi_k \leq \frac{k}{2^m}$$

and

$$E\mathbf{x} = \lim_{n\to\infty} \sum_{k=-\infty}^{\infty} \frac{k-1}{2^n}\left[F\left(\frac{k}{2^n}\right) - F\left(\frac{k-1}{2^n}\right)\right].$$

Finally, from the above expressions for S and T, it is clear that $E\mathbf{x}$ exists if $E|\mathbf{x}|$ does, and in fact $E|\mathbf{x}| = S + T$.

Next let us show that there is no disagreement between the Lebesgue definition of expectation and the two previous definitions given in Chapters 3 and 4, respectively. That is, the Lebesgue definition yields the others as special cases.

Theorem 8.1

If $F(x)$ is an absolutely continuous d.f. then

$$\int \mathbf{x}\,dP = \int_{-\infty}^{\infty} xf(x)\,dx. \tag{8.2}$$

Hence, the Lebesgue expectation agrees with that of Chapter 4.

Proof

We must show that when one side of (8.2) exists then so does the other and the two are equal. We have

$$\int_0^n xf(x)\,dx - \frac{1}{2^n} \le \int_0^n \left(x - \frac{1}{2^n}\right) f(x)\,dx = \sum_{k=1}^{n2^n} \int_{(k-1)/2^n}^{k/2^n} \left(x - \frac{1}{2^n}\right) f(x)\,dx$$

$$\le \sum_{k=1}^{n2^n} \frac{k-1}{2^n} \int_{(k-1)/2^n}^{k/2^n} f(x)\,dx = S_n - n[1 - F(n)]$$

and similarly

$$S_n - n[1 - F(n)] \le \int_0^n xf(x)\,dx$$

so that

$$0 \le \int_0^n xf(x)\,dx + n[1 - F(n)] - S_n \le \frac{1}{2^n}.$$

From this pair of inequalities we see that $S = \int_0^\infty xf(x)\,dx$, if $n[1 - F(n)] \to 0$. But we have already argued that $n[1 - F(n)] \to 0$, if $S < \infty$. On the other hand, if $\int_0^\infty xf(x)\,dx$ converges then

$$0 \le n[1 - F(n)] = n\int_n^\infty f(x)\,dx \le \int_n^\infty xf(x)\,dx \to 0.$$

Thus, in summary, $n[1 - F(n)] \to 0$ and $S = \int_0^\infty xf(x)\,dx$, if either $S < \infty$ or $\int_0^\infty xf(x)\,dx < \infty$. Hence $S < \infty$ iff $\int_0^\infty xf(x)\,dx < \infty$ and if either is finite, the two are equal. A similar argument involving T proves the theorem.

Theorem 8.2

If x is a discrete r.v. then

$$\int \mathbf{x}\,dP = \sum_X xp(x).$$

Proof

For $\mathbf{x}$ discrete

$$S_n = \sum_{0<x\le n} \frac{\langle x \cdot 2^n \rangle}{2^n} p(x) + n[1 - F(n)]$$

where $\langle y \rangle$ is the greatest integer *less* than y. Then

$$\sum_{0<x\leq n} xp(x) + n[1 - F(n)] - S_n = \sum_{0<x\leq n} \left(x - \frac{\langle x \cdot 2^n \rangle}{2^n}\right) p(x)$$

and

$$0 < \sum_{0<x\leq n} xp(x) + n[1 - F(n)] - S_n \leq \frac{1}{2^n}.$$

Now if $\sum_{0<x} xp(x) < \infty$ then

$$n[1 - F(n)] = n \sum_{n<x} p(x) < \sum_{n<x} xp(x) \to 0.$$

We see that the proof follows the lines of the previous theorem.

With the Lebesgue concept of expectation, Theorem 3.1 generalizes to Theorem 8.3.

Theorem 8.3

If $F_{\mathbf{x}}(x)$ is the d.f. of the r.v. $\mathbf{x}$ and $g(x)$ is a continuous function, then

$$Eg(\mathbf{x}) = \int g(x)\, dF_{\mathbf{x}}(x).$$

Theorem 3.2 remains valid as it is with the generalized concept of expectation.

8.2 SOME THEORETICAL EXAMPLES

The d.f.'s treated up to this point tend to leave the reader with a general feeling of comfort and well-being concerning the behavior of distribution functions. Distributions are either discrete or smooth, moments exist, $\bar{\mathbf{x}}$ approaches its expectation in probability, and for large n, $\bar{\mathbf{x}}$ tends to be normally distributed with ever decreasing dispersion. While this general point of view is by and large to be cultivated, the present section is included as an antidote to over-optimism.

First consider the relation between d.f. $F(x)$ and density $f(x)$ for absolutely continuous d.f.'s. In analysis, beyond the scope of this writing, it is shown that $f(x) = F'(x)$ except on a set of combined length zero. For example, if $f(x) = 1$ for irrational x between 0 and 1 and $f(x) = 0$ elsewhere on the real line, then the Lebesgue integral $\int_{-\infty}^{x} f(y)\, dy$ is the uniform distribution of Equation (4.5) and $F'(x) = 1$ or 0 according as x is or is not in the interval (0, 1). Thus, $f(x) = F'(x)$, except for rational x on the unit interval. But these rationals must have combined length zero since they are countable, and hence, can be covered by a countable collection of intervals having total length ϵ. (The first rational in some listing is covered by an interval of length $\epsilon/2$, the second by an interval of length $\epsilon/2^2$, and so forth.)

Finally we treat two well-known distributions which "behave badly."

The *Cauchy distribution* with density

$$c(x; \lambda, \mu) = \frac{1}{\pi} \frac{\lambda}{\lambda^2 + (x - \mu)^2}, \quad \lambda > 0$$

is smooth, "bell shaped," and symmetric about μ, very much like the normal. The points of inflection are $\mu \pm \lambda/\sqrt{3}$ so that λ is a measure of dispersion much like the standard deviation of the normal. And yet despite this bell-shaped appearance, no positive moment, not even the mean, exists. For example, with $\mu = 0$ and $\lambda = 1$, S of Section 8.1 is given by

$$S = \lim_{u \to \infty} \int_0^u \frac{x\,dx}{\pi(1 + x^2)} = \lim_{u \to \infty} \frac{1}{2\pi} \log(1 + u^2) = \infty.$$

Similarly, $T = \infty$ and the mean does not exist.

The c.f. of $c(x; 1, 0)$ is

$$\int_{-\infty}^{\infty} e^{itx} c(x; 1, 0)\,dx = 2 \int_0^{\infty} \cos tx \cdot c(x; 1, 0)\,dx = \exp(-|t|)$$

so that the c.f. of $c(x; \lambda, \mu)$ is $\exp(it\mu - \lambda|t|)$.

We are now in a position to examine the asymptotic behavior of the distribution of $\bar{\mathbf{x}}$, the average of n independent Cauchy r.v.'s. The c.f. of $\bar{\mathbf{x}}$ is

$$\left\{\exp\left[i\left(\frac{t}{n}\right)\mu - \lambda\left|\frac{t}{n}\right|\right]\right\}^n = \exp(it\mu - \lambda|t|).$$

$\bar{\mathbf{x}}$ has the same Cauchy distribution as one of the variables being averaged; there is no convergence of $\bar{\mathbf{x}}$ to its expectation and no convergence toward normality.

The above property uniquely characterizes the Cauchy distribution.

Theorem 8.4

If $F(x)$ is a nondegenerate d.f. having the property that the mean of a sample also has d.f. $F(x)$, then $F(x)$ is a Cauchy distribution.

Proof

Denoting by $\phi(t)$ the c.f. of F, we may restate our problem as that of solving the functional equation

$$\left[\phi\left(\frac{t}{n}\right)\right]^n = \phi(t).$$

For positive integral m and n we obtain

$$\phi\left(\frac{m}{n}\right) = [\phi(m)]^{1/n} = [\phi(1)]^{m/n},$$

and hence, by continuity $\phi(t) = [\phi(1)]^t = (\rho e^{i\theta})^t$ for any real $t \geq 0$. But $\phi(-t) = \overline{\phi(t)}$ so that $\phi(t) = \rho^{|t|}e^{it\theta}$ for t real. Since ϕ is bounded, we may write $\rho = e^{-\lambda}$ where $\lambda \geq 0$. The resulting characteristic function, $\phi(t) = \exp(-\lambda|t| + it\theta)$, is that of a degenerate or a Cauchy distribution according as $\lambda = 0$ or $\lambda > 0$.

We now turn to the *Cantor distribution* which can be given the following intuitive background. Let $\mathbf{x}$ be the gain of a gambler who receives the amount $2 \cdot 3^{-k}$, if the kth in an infinite sequence of tosses of a fair coin results in heads. Thus

$$\mathbf{x} = \sum_{k=1}^{\infty} \frac{\mathbf{x}_k}{3^k},$$

where $\mathbf{x}_k = 2$ or 0 according as the kth trial results in head or tail. What is the distribution of $\mathbf{x}$?

First we recall that every point a on the interval [0, 1] can be written in the form

$$a = \sum_{k=1}^{\infty} \frac{a_k}{3^k},$$

that is, in triadic expansion as $a = .a_1a_2\cdots$, where $a_k = 0, 1, 2$, for each k. Two numbers a and b $(= .b_1b_2\cdots)$ may be compared by examining their triadic expansions. If $a_i = b_i$, for $i = 1, 2, \ldots,$ then of course $a = b$; if $a_i = b_i$, for $i = 1, 2, \ldots, k-1$, but $a_k < b_k$, then

$$a \leq .a_1 \cdots a_{k-1}b_k0 \cdots \leq .b_1b_2 \cdots = b.$$

Hence, since $P(\mathbf{x}_i = a_i \text{ for } i = 1, 2, \ldots) = 0$,

$$F(a) = P(\mathbf{x} \leq a) = P(\mathbf{x}_1 < a_1) + P(\mathbf{x}_1 = a_1, \mathbf{x}_2 < a_2) + P(\mathbf{x}_1 = a_1, \mathbf{x}_2 = a_2, \mathbf{x}_3 < a_3) + \cdots.$$

If $a = .a_1 \cdots a_{k-1}1a_{k+1}\cdots$ with $a_i \neq 1$ for $i = 1, \ldots, k-1$, then

$$F(a) = \frac{1}{2}\left(\frac{a_1}{2} + \cdots + \frac{a_{k-1}}{2^{k-1}}\right) + \frac{1}{2^k}.$$

On the other hand, if a has no 1's in its expansion, then

$$F(a) = \frac{1}{2}\sum_{i=1}^{\infty} \frac{a_i}{2^i}.$$

From this expression for $F(x)$, it is clear that $F(x) \to F(a)$ as $x \to a$, that is, $F(x)$ is a continuous function.

The Cantor distribution has the form suggested by Figure 8.3. There is a single interval of constancy of length $\frac{1}{3}$, 2 intervals of length $\frac{1}{9}$, 4 of length $\frac{1}{27}$, and so on. The total length of all intervals of constancy is

$$\frac{1}{3} + \frac{2}{3^2} + \frac{2^2}{3^3} + \cdots = 1.$$

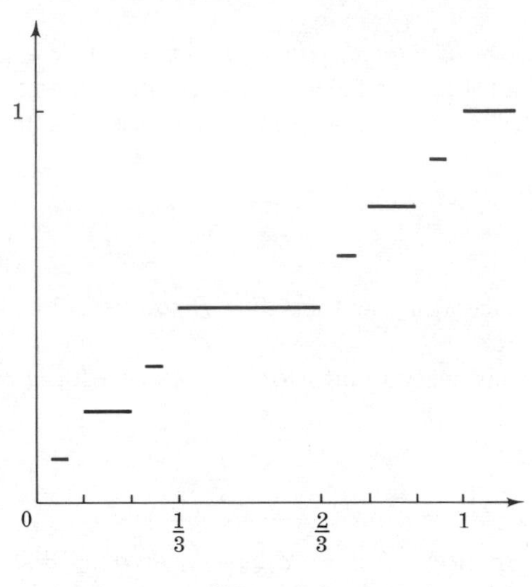

Figure 8.3

F increases from 0 to 1 in such a way that the total increase takes place on a set of zero length.

F does not have a density function, it is not representable in the form

$$F(x) = \int_{-\infty}^{x} f(t)\, dt.$$

From analysis, if such a representation were possible then we would have $F'(x) = f(x)$ except on a set of zero length. But $F'(x) = 0$, so $f(x) = 0$, except on a set of zero length. Then

$$\int_{-\infty}^{\infty} f(x)\, dx = 0,$$

which is a contradiction.

Thus, the Cantor distribution is an example of a continuous d.f. which is not absolutely continuous: it does not have a density function.

PROBLEMS

1. Using (8.1), calculate the expectation of an r.v. having the uniform distribution on [0, 1].

2. Let $\mathbf{x}$ have d.f.

$$F(x) = \begin{cases} 0, & x < e \\ 1 - (x \log x)^{-1}, & x \geq e. \end{cases}$$

$\lim_{n \to \infty} n[1 - F(n)] = 0$? Does $E\mathbf{x}$ exist?

3. Show that if

$$\mathbf{x} = \sum_{k=1}^{\infty} \frac{\mathbf{x}_k}{2^k}$$

where the $\mathbf{x}_k$ are independent r.v.'s assuming values 0 and 1 with equal probability, then $\mathbf{x}_k$ has a uniform d.f.

4. Prove in general (or else for the continuous and discrete cases) that if $\mathbf{x}$ has a symmetric distribution about c, that is, $F(c + x) = 1 - F(c - x)$, then $E\mathbf{x} = c$.

5. Evaluate the moment generating function of the Cauchy distribution.

9

MULTIVARIATE DISTRIBUTIONS

9.1 GENERAL MULTIVARIATE DISTRIBUTIONS

We have discussed joint, marginal, and conditional probabilities for two experiments, but straightforward generalization is possible to an arbitrary finite number. We may discuss r.v.'s $\mathbf{y}_1, \ldots, \mathbf{y}_p$ with d.f.

$$G(y_1, \ldots, y_p) = P(\mathbf{y}_1 \leq y_1, \ldots, \mathbf{y}_p \leq y_p),$$

and (if it exists) density function $g(y_1, \ldots, y_p)$ satisfying

$$G(y_1, \ldots, y_p) = \int_{-\infty}^{y_p} \cdots \int_{-\infty}^{y_1} g(r_1, \ldots, r_p)\, dr_1 \cdots dr_p.$$

Marginal distributions and densities are obtained from their joint counterparts by formulas generalizing Equations (4.12) and (4.13). The marginal density of any subset of r.v.'s is obtained by "integrating out" the others. Conditional distributions and densities are defined in a manner entirely analogous to the two-dimensional case. For example, with three variables $\mathbf{x}$, $\mathbf{y}$, and $\mathbf{z}$

$$G_{\mathbf{x},\mathbf{y}}(x, y) = \lim_{z \to \infty} G(x, y, z)$$

and, assuming a density function,

$$g_{\mathbf{x}}(x) = \int_{-\infty}^{\infty} \int_{-\infty}^{\infty} g(x, s, t)\, ds\, dt.$$

The conditional density of $\mathbf{x}$ given that $\mathbf{y} = y$ and $\mathbf{z} = z$ is

$$\frac{g(x, y, z)}{g_{\mathbf{y},\mathbf{z}}(y, z)} = \frac{g(x, y, z)}{\int_{-\infty}^{\infty} g(t, y, z)\, dt}.$$

Variables **x**, **y**, and **z** are independent if

$$G(x, y, z) = G_{\mathbf{x}}(x) \cdot G_{\mathbf{y}}(y) \cdot G_{\mathbf{z}}(z).$$

The joint c.f. of $\mathbf{y}_1, \ldots, \mathbf{y}_p$ is

$$\phi_{\mathbf{y}_1, \ldots, \mathbf{y}_p}(t_1, \ldots, t_p) = E \exp i(t_1\mathbf{y}_1 + \cdots + t_p\mathbf{y}_p).$$

Uniqueness and continuity results generalizing Theorems 7.9 and 7.10 hold for joint characteristic functions and there is a multivariate central limit theorem analogous to Theorem 7.11. Using the uniqueness theorem, it can be shown that r.v.'s are independent if and only if their individual c.f.'s multiply to give the joint c.f.

In more dimensions, transformation of variables is handled in the same way as in Section 4.5. That is, if $\mathbf{y}_i = u_i(\mathbf{x}_1, \ldots, \mathbf{x}_p)$; $i = 1, \ldots, p$ and if $g(x_1, \ldots, x_p)$ is the density of $\mathbf{x}_1, \ldots, \mathbf{x}_p$ then, under regularity conditions,

$$g(y_1, \ldots, y_p) = f[x_1(y_1, \ldots, y_p), \ldots, x_p(y_1, \ldots, y_p)] \cdot J$$

where $x_i(y_1, \ldots, y_p)$ is the inverse transformation of $y_i = u_i(x_1, \ldots, x_p)$ and J, the Jacobian, is the absolute value of the determinant

$$\begin{vmatrix} \dfrac{\partial x_1}{\partial y_1} & \cdots & \dfrac{\partial x_p}{\partial y_1} \\ \vdots & & \vdots \\ \dfrac{\partial x_1}{\partial y_p} & \cdots & \dfrac{\partial x_p}{\partial y_p} \end{vmatrix}$$

9.2 THE MULTIVARIATE NORMAL DISTRIBUTION

Before beginning this section the reader may wish to refresh his memory on vectors, matrices, and determinants by reading the Appendix on these subjects. Here, we take as our starting point the following definition. If $\mathbf{x}_1, \ldots, \mathbf{x}_q$ are independent $N(0, 1)$ and

$$\begin{aligned} \mathbf{y}_1 &= a_{11}\mathbf{x}_1 + \cdots + a_{1q}\mathbf{x}_q + \mu_1 \\ \mathbf{y}_2 &= a_{21}\mathbf{x}_1 + \cdots + a_{2q}\mathbf{x}_q + \mu_2 \\ \vdots & \quad \vdots \quad \quad \quad \quad \vdots \\ \mathbf{y}_p &= a_{p1}\mathbf{x}_1 + \cdots + a_{pq}\mathbf{x}_q + \mu_p, \end{aligned}$$

then $\mathbf{y}_1, \mathbf{y}_2, \ldots, \mathbf{y}_p$ have the *p- or multivariate normal* (*MVN*) *distribution.* From this definition it is immediately clear that linear combinations of MVN random variables are MVN. We may easily compute

$$E\mathbf{y}_i = \mu_i, \quad V(\mathbf{y}_i) = \sum_{j=1}^{q} a_{ij}^2$$

and

$$\text{Cov}(\mathbf{y}_i, \mathbf{y}_{i'}) = \sum_{j=1}^{q} a_{ij} \cdot a_{i'j}.$$

Or in vector notation if $\mathbf{y}^T = (\mathbf{y}_1, \ldots, \mathbf{y}_p)$, $\mathbf{x}^T = (\mathbf{x}_1, \ldots, \mathbf{x}_q)$, $\mu^T = (\mu_1, \ldots, \mu_p)$, $A = (a_{ij})$ and $\mathbf{y} = A\mathbf{x} + \mu$ then the expectation vector and covariance matrix of $\mathbf{y}$ are

$$E\mathbf{y} = \mu \quad \text{and} \quad \Sigma_{\mathbf{y}} = AA^T.$$

Theorem 9.1

The joint characteristic function of the MVN is

$$\phi(t_1, \ldots, t_p) = \exp(it^T\mu - \tfrac{1}{2}t^T\Sigma_{\mathbf{y}}t).$$

Proof

By definition

$$\begin{aligned}\phi(t_1, \ldots, t_p) &= Ee^{i(t_1\mathbf{y}_1 + \cdots + t_p\mathbf{y}_p)} = Ee^{it^T\mathbf{y}} \\ &= Ee^{it^T(A\mathbf{x} + \mu)} \\ &= e^{it^T\mu} \cdot Ee^{it^TA\mathbf{x}}.\end{aligned}$$

Now writing $s^T = t^TA$, and using the independence of the $\mathbf{x}$'s,

$$\begin{aligned}Ee^{it^TA\mathbf{x}} &= Ee^{is^T\mathbf{x}} = \prod_{j=1}^{q} Ee^{is_j\mathbf{x}_j} \\ &= e^{-(s_1{}^2 + \cdots + s_q{}^2)/2} = e^{-s^Ts/2} \\ &= e^{-t^TAA^Tt/2} = e^{-t^T\Sigma_{\mathbf{y}}t/2}.\end{aligned}$$

The joint characteristic function is one way of representing the MVN distribution. We see that the distribution depends on A only through $\Sigma_{\mathbf{y}} = AA^T$ so that several A matrices could lead to the same distribution of $\mathbf{y}$. The distribution is determined by the moments of first and second order. Accordingly we write: $\mathbf{y}$ is $N(\mu, \Sigma_{\mathbf{y}})$ to mean that $\mathbf{y}$ has the MVN distribution with mean vector μ and variance-covariance matrix Σ_y.

Remember from Section 7.1 that (i) the correlation coefficient of arbitrary r.v.'s is between 1 and -1, and (ii) independence implies uncorrelated variables. For the MVN, the correlation coefficient

$$R(\mathbf{y}_i, \mathbf{y}_{i'}) = \frac{\sum_j a_{ij}a_{i'j}}{\sqrt{(\sum_j a_{ij}{}^2)(\sum_j a_{i'j}{}^2)}}$$

can be seen to assume all values throughout its range including the endpoints. Also the r.v.'s $\mathbf{y}_i$ and $\mathbf{y}_{i'}$ are uncorrelated if and only if the vectors $(a_{i1}, \ldots, a_{iq})$ and $(a_{i'1}, \ldots, a_{i'q})$ are orthogonal, but these are exactly the circumstances under which the characteristic function given in Theorem 9.1 will factor into a product of characteristic functions of the individual r.v's. Hencc, for jointly normal r.v.'s, independence is equivalent to being uncorrelated.

The next result is immediately apparent from the characteristic function representation (and indeed from the definition itself).

Theorem 9.2

The marginal distribution of $\mathbf{y}_1, \ldots, \mathbf{y}_k$ $(k < p)$ is $N(\mu^*, \Sigma^*)$ where μ^* consists of the first k elements of μ and Σ^* is the $k \times k$ matrix in the upper left-hand corner of $\Sigma_{\mathbf{y}}$. If other than the first k variables are involved, one permutes the elements of $\mathbf{y}$ and proceeds as above.

A second method of representation, when it exists, is the joint density function.

Theorem 9.3

If $\Sigma_{\mathbf{y}}^{-1}$ exists, the joint density function of $\mathbf{y}_1, \ldots, \mathbf{y}_p$ is given by

$$f(y_1, \ldots, y_p) = (2\pi)^{p/2}|\Sigma_{\mathbf{y}}|^{-1/2} \exp\left[-\tfrac{1}{2}(y - \mu)^T\Sigma_{\mathbf{y}}^{-1}(y - \mu)\right].$$

Proof

Choose $q = p$ with A nonsingular and such that $AA^T = \Sigma_{\mathbf{y}}$. Since A is nonsingular, A^{-1} exists. The joint density function of $\mathbf{x}_1, \ldots, \mathbf{x}_p$ is

$$(2\pi)^{-p/2} \exp\left(-\tfrac{1}{2}x^Tx\right).$$

Making the transformation $\mathbf{y} = Ax + \mu$ or $\mathbf{x} = A^{-1}(\mathbf{y} - \mu)$ with Jacobian $|A^{-1}| = |A|^{-1}$ we find that the density of $\mathbf{y}$ is

$$\frac{1}{(2\pi)^{p/2}|A|} e^{-(y-\mu)^T(A^{-1})^TA^{-1}(y-\mu)/2}.$$

But $|\Sigma_{\mathbf{y}}| = |A|\,|A^T| = |A|^2$, so that $|A| = |\Sigma_{\mathbf{y}}|^{1/2}$ and $(A^{-1})^TA^{-1} = \Sigma_{\mathbf{y}}^{-1}$ as the following computation shows.

$$\begin{aligned}\Sigma_{\mathbf{y}}(A^{-1})^T(A^{-1}) &= AA^T(A^{-1})^TA^{-1} = A(A^{-1}A)^TA^{-1} \\ &= AIA^{-1} = I.\end{aligned}$$

This gives the equation of the theorem.

Note that for the MVN, the contours of equal probability density are similar ellipsoids all having center μ. A "contour map" of the bivariate normal density would have the general appearance as shown in Figure 9.1.

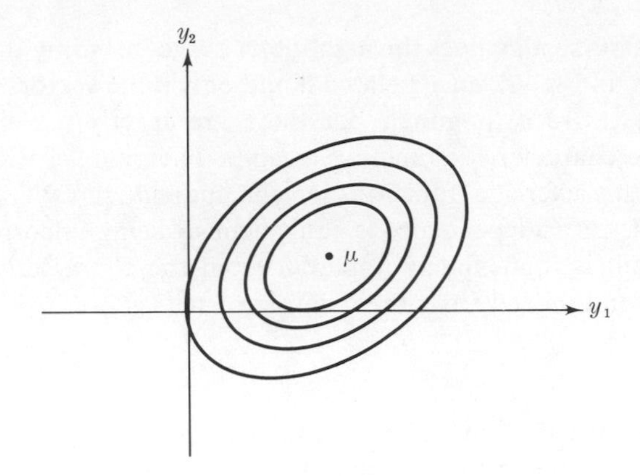

Figure 9.1

9.3 SAMPLING DISTRIBUTIONS ASSOCIATED WITH THE NORMAL

In statistics, sampling from a prescribed distribution is an important concept. A sample of size n from the univariate normal distribution, mean μ and variance σ^2, consists of n independent identically distributed r.v.'s, $\mathbf{x}_1, \ldots, \mathbf{x}_n$, all having density $n(x; \mu, \sigma^2)$.

If $\mu = 0$ and $\sigma^2 = 1$, then we say that $\boldsymbol{\chi}_n{}^2 = \mathbf{x}_1{}^2 + \cdots + \mathbf{x}_n{}^2$ has the *chi-square* distribution with n degrees of freedom. By direct computation, $\mathbf{x}_i{}^2$ has the density $\gamma(x; \frac{1}{2}, \frac{1}{2})$ of Section 4.7 so that from Theorem 4.2 we see that $\boldsymbol{\chi}_n{}^2$ has the special gamma density $\gamma(x; n/2, 1/2)$.

Further, if $\mathbf{x}, \mathbf{x}_1, \ldots, \mathbf{x}_n$ are independent $N(0, \sigma^2)$, then

$$\mathbf{t} = \frac{\mathbf{x}}{\sqrt{\dfrac{\sum \mathbf{x}_i{}^2}{n}}} = \frac{\mathbf{x}}{\sqrt{\dfrac{\boldsymbol{\chi}_n{}^2}{n}}}$$

has *Student's t* distribution with n degrees of freedom. To find the density of $\mathbf{t}$, we apply Theorem 4.2 once again. From that theorem $\mathbf{w} = \mathbf{x}^2/(\mathbf{x}^2 + \boldsymbol{\chi}_n{}^2)$ has density

$$\beta\left(w; \frac{1}{2}, \frac{n}{2}\right) = \frac{\Gamma\left(\dfrac{n+1}{2}\right)}{\Gamma\left(\dfrac{1}{2}\right)\Gamma\left(\dfrac{n}{2}\right)} w^{-1/2}(1 - w)^{(n/2)-1}$$

but $\mathbf{w} = (1 + n/\mathbf{t}^2)^{-1}$ so that $\mathbf{t}$ has density

$$\frac{1}{\sqrt{n\pi}} \frac{\Gamma\left(\dfrac{n+1}{2}\right)}{\Gamma\left(\dfrac{n}{2}\right)} \left(1 + \frac{t^2}{n}\right)^{-(n+1)/2}$$

The basic result concerning sampling from the univariate normal is the following.

Theorem 9.4

If $\mathbf{x}_1, \ldots, \mathbf{x}_n$ constitute a sample from the density $n(x; \mu, \sigma^2)$, then

(i) $\bar{\mathbf{x}} = \sum_i \mathbf{x}_i/n$ has density $n(x; \mu, \sigma^2/n)$,
(ii) $\sum_i (\mathbf{x}_i - \bar{\mathbf{x}})^2/\sigma^2$ has the chi-square density with $n - 1$ degrees of freedom,
(iii) $\bar{\mathbf{x}}$ and $\sum_i (\mathbf{x}_i - \bar{\mathbf{x}})^2$ are independently distributed,

and

(iv)
$$\frac{\sqrt{n}(\bar{\mathbf{x}} - \mu)}{\sqrt{\dfrac{\sum(\mathbf{x}_i - \bar{\mathbf{x}})^2}{n - 1}}}$$

has the t distribution with $n - 1$ degrees of freedom.

Proof

Property (iv) will be clear from the definition of the t distribution as soon as (i), (ii), and (iii) are proved. The joint density function of $\mathbf{x}_1, \ldots, \mathbf{x}_n$ is

$$(2\pi\sigma^2)^{-n/2} \exp\left\{-\frac{\sum_1^n (x_i - \mu)^2}{2\sigma^2}\right\}.$$

Perform the orthogonal transformation $y = Ax$, where

$$A = \begin{pmatrix} \frac{1}{\sqrt{2}} & \frac{-1}{\sqrt{2}} & 0 & \cdots & 0 \\ \frac{1}{\sqrt{6}} & \frac{1}{\sqrt{6}} & \frac{-2}{\sqrt{6}} & \cdots & 0 \\ \vdots & \vdots & \vdots & & \vdots \\ \frac{1}{\sqrt{n(n-1)}} & \frac{1}{\sqrt{n(n-1)}} & \frac{1}{\sqrt{n(n-1)}} & & \frac{-(n-1)}{\sqrt{n(n-1)}} \\ \frac{1}{\sqrt{n}} & \frac{1}{\sqrt{n}} & \frac{1}{\sqrt{n}} & \cdots & \frac{1}{\sqrt{n}} \end{pmatrix},$$

the so-called *Helmert matrix*. The Jacobian is 1 and

$$\begin{aligned}\sum_1^n (x_i - \mu)^2 &= \sum_1^n x_i^2 - n\bar{x}^2 + n(\bar{x} - \mu)^2 \\ &= \sum_1^n y_i^2 - y_n^2 + n\left(\frac{y_n}{\sqrt{n}} - \mu\right)^2.\end{aligned}$$

Hence, the joint density of $\mathbf{y}_1, \ldots, \mathbf{y}_n$ is

$$(2\pi\sigma^2)^{-n/2} \exp\left\{-\frac{1}{2\sigma^2}\left[\sum_1^{n-1} y_i^2 + n\left(\frac{y_n}{\sqrt{n}} - \mu\right)^2\right]\right\},$$

that is, $\mathbf{y}_1, \ldots, \mathbf{y}_n$ are independent and normal with common variance σ^2. Properties (i) and (ii) follow from the identities $\mathbf{y}_n = \sqrt{n}\,\bar{\mathbf{x}}$ and

$$\sum_1^{n-1} \mathbf{y}_i^2 = \sum_1^{n} (\mathbf{x}_i - \bar{\mathbf{x}})^2,$$

property (iii) follows from the independence of the $\mathbf{y}$'s.

9.4 AN APPROXIMATION TO THE DISTRIBUTION OF A QUADRATIC FORM

In many applied problems, one becomes interested in the distribution of a quadratic form in normal random variables. For example, this question arises when one considers a rifleman practice-firing on a circular target. Vertical and lateral miss distance may be considered to be bivariate normal in distribution and the rifleman's score will be determined by radial miss distance from the center of the target.

In the most general case, the quantity being approximated may be written explicitly as

$$P(\mathbf{Q} \leq t) = \text{const.} \int_{Q \leq t} \cdots \int \exp\left(-\tfrac{1}{2}\sum_{i,j=1}^{p} \lambda_{ij} y_i y_j\right) dy_1 \cdots dy_p$$

where

$$Q = \sum_{i,j=1}^{p} a_{ij} y_i y_j$$

is a positive definite quadratic form and $\lambda = (\lambda_{ij})$ is the inverse of the covariance matrix $\Sigma = (\sigma_{ij})$. This integral can be transformed[1] to obtain the simpler expression

$$P(\mathbf{Q} \leq t) = c \int_{\sum_i y_i^2 \leq t} \cdots \int \exp\left(-\frac{1}{2}\sum_i \frac{y_i^2}{a_i}\right) dy_1 \cdots dy_p \tag{9.1}$$

where $a_1, \ldots, a_p$ are the (necessarily positive) roots of $|A\Sigma - aI| = 0$ and

$$c^{-1} = (2\pi)^{p/2} \prod_1^p a_i^{1/2}.$$

[1] Anderson, T. W., *An Introduction to Multivariate Statistical Analysis.* Wiley, New York, 1958, p. 341.

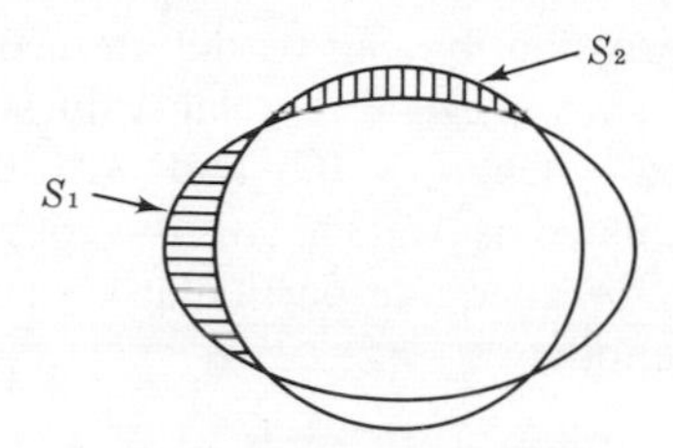

Figure 9.2

As a preliminary, we derive an easy inequality for the expression (9.1). In two dimensions, the regions making equal contributions per unit of area are elliptic shells. Thus, if we consider integrating over the interiors of an ellipse and a circle of equal area the contribution of the area S_1, shown in Figure 9.2, is greater than the contribution of S_2.

A similar argument could be presented in p-dimensions. Thus,

$$\int_{\Sigma y_i^2 \le t} \cdots \int \exp\left(-\frac{1}{2}\sum_i \frac{y_i^2}{a_i}\right) dy_1 \cdots dy_p$$

$$< \int_{\Sigma_i y_i^2/a_i \le t/(\Pi a_i)^{1/p}} \cdots \int \exp\left(-\frac{1}{2}\sum \frac{y_i^2}{a_i}\right) dy_1 \cdots dy_p.$$

But this last integral, except for the constant c, is the chi-square distribution with p degrees of freedom. Hence,

$$P(\mathbf{Q} \le t) < P\left[\chi_p^2 \le \frac{t}{(\prod_i a_i)^{1/p}}\right] \tag{9.2}$$

with equality holding when the a's are equal and inequality becoming greater as the a's depart more from one another. We conclude that the chi-square approximation of (9.2), based on the geometric mean of the a's, leads to an overestimate.

On the other hand, we have the underestimate

$$P\left(\chi_p^2 \le \frac{t}{\max\{a_i\}}\right) < P(\mathbf{Q} \le t).$$

Since $\max\{a_i\} > \bar{a} > (\prod_i a_i)^{1/p}$, we tentatively suggest that an approximation similar to (9.2) but based on $\bar{a}$ instead of $(\prod_i a_i)^{1/p}$ may be about right. This choice, among rough approximations of the type (9.2), is particularly appealing because of its algebraic simplicity; we have $\bar{a} = \sum_i a_{ii}/p$ if $\Sigma = I$, $\bar{a} = \sum_i \sigma_{ii}/p$ if $A = I$, and $\bar{a} = \sum_{i,j} a_{ij}\sigma_{ij}/p$ in general.

In two and three dimensions, the most likely cases of practical interest, the error of the $\bar{a}$ approximation can be determined from published tables.[2] For convenience we show several typical values, for example, if the $\bar{a}$ approximation is used when $a_2 = 0.8$ and $a_1 = 0.2$, then $P(\mathbf{Q} \leq 0.7) = 0.5464$ will be approximated as 0.5034. This is a rough approximation but for some purposes it may be adequate particularly when the a's do not depart very much from one another.

$P(\mathbf{Q} \leq t)$ in two dimensions

t	$\bar{a}$ *approximation*	a_2, a_1			
		0.6, 0.4	0.8, 0.2	0.9, 0.1	0.99, 0.01
0.7	0.5034	0.5080	0.5464	0.5780	0.5962
3.0	0.9502	0.9487	0.9365	0.9269	0.9178

$P(\mathbf{Q} \leq t)$ in three dimensions

t	$\bar{a}$ *approximation*	a_3, a_2, a_1		
		0.4, 0.4, 0.2	0.6, 0.2, 0.2	0.8, 0.1, 0.1
0.7	0.50637	0.5161	0.5402	0.5974
3.0	0.97071	0.9668	0.9577	0.9378

9.5 APPLICATION—COMPONENT TOLERANCES WHICH ACHIEVE A SPECIFIED SYSTEM TOLERANCE

Here we treat one specific aspect of a wider problem of considerable practical interest and significance: How should component variabilities (or tolerances) be chosen in order to stay within a specified system variability? The $\bar{a}$ approximation of Section 9.4 is used to answer this question for the multivariate normal case when the quantity of interest is Euclidean distance.

As a simple example, we consider rifle marksmanship in greater detail. Here we consider a two component system (i) the marksman with his weapon and (ii) the ammunition. Let $\mathbf{X}_1 = (\mathbf{x}_1, \mathbf{y}_1)$ be the two dimensional aiming error vector of the marksman and let $\mathbf{X}_2 = (\mathbf{x}_2, \mathbf{y}_2)$ be the vector of ammunition dispersion error. $\mathbf{X} = (\mathbf{x}, \mathbf{y}) = (\mathbf{x}_1 + \mathbf{x}_2, \mathbf{y}_1 + \mathbf{y}_2)$ is then the total error vector of the system. We assume the components $\mathbf{X}_1$ and $\mathbf{X}_2$ to be independent and bivariate normal with zero means. Let σ_x^2, σ_y^2, and σ_{xy}

[2] Grad, Arthur and Herbert Solomon, Distribution of quadratic forms and some applications. *Ann. Math. Statist.*, **26**, 464–477 (1955).

be the variances and covariances of the horizontal and vertical components of error due to the marksman's aim or skill, and let τ_x^2, τ_y^2, and τ_{xy} be similarly defined for the dispersion error of the ammunition. Then $\Sigma = \Sigma_1 + \Sigma_2$ is the variance-covariance matrix of $\mathbf{X}$ where

$$\Sigma_1 = \begin{pmatrix} \sigma_x^2 & \sigma_{xy} \\ \sigma_{xy} & \sigma_y^2 \end{pmatrix} \quad \text{and} \quad \Sigma_2 = \begin{pmatrix} \tau_x^2 & \tau_{xy} \\ \tau_{xy} & \tau_y^2 \end{pmatrix}.$$

The specific problem under consideration is that of determining how small these errors must be in order that the bullet impact falls within a circle of specified radius with some assigned probability.

Using the $\bar{a}$ approximation with $A = I$ we have

$$P(\mathbf{Q} \leq t) \simeq P\left(\chi_2^2 \leq \frac{t}{\bar{a}}\right)$$

where

$$\bar{a} = \frac{\sigma_{11} + \sigma_{22}}{2} = \frac{(\sigma_x^2 + \tau_x^2) + (\sigma_y^2 + \tau_y^2)}{2}.$$

For example, impact will fall within a radius, say two feet, with a probability of approximately $\frac{1}{2}$ whenever

$$\sigma_x^2 + \sigma_y^2 + \tau_x^2 + \tau_y^2 \leq \frac{2 \cdot 2^2}{1.386} \simeq 5.8.$$

9.6 THE MULTIVARIATE EXPONENTIAL DISTRIBUTION[3]

In previous chapters we have emphasized the basic importance of the exponential distribution. A generalization to higher dimensions should prove equally fundamental in handling more variables. $\mathbf{x}$ and $\mathbf{y}$ have the *bivariate exponential* (BVE) distribution, if for $s, t > 0$,

$$\begin{aligned} \bar{F}(s, t) &= P(\mathbf{x} > s, \mathbf{y} > t) \\ &= \exp\left[-\lambda_1 s - \lambda_2 t - \lambda_{12} \max(s, t)\right] \end{aligned} \tag{9.3}$$

where $\lambda_1, \lambda_2, \lambda_{12} \geq 0$.

We will derive the expression (9.3) from three separate models. The first is a "fatal shock" model. Suppose that the components of a two-component system invariably fail after receiving a shock. Shocks of three types arrive independently and at random in time. Shocks of the first type cause the first component to fail and have expected rate of arrival λ_1; those

[3] Marshall, Albert W., and Ingram Olkin, A multivariate exponential distribution. *J. Amer. Statist. Assn.*, **62**, 30–44 (1967).

of the second type cause the second component to fail and have expected rate of arrival λ_2. The third type of shock causes both components to fail; its expected rate of arrival is λ_{12}. If **x** and **y** are the survival times of the first and second components then

$$\begin{aligned}\bar{F}(s, t) = &P(\text{no type 1 arrivals in time } s) \\ &\cdot P(\text{no type 2 arrivals in time } t) \\ &\cdot P(\text{no type 3 arrivals in time } \max(s, t))\end{aligned}$$

which gives Equation (9.3).

An immediate and useful consequence of the fatal shock model is Theorem 9.5.

Theorem 9.5

x and **y** are BVE if and only if there exist independent exponential r.v.'s **u**, **v**, and **w** such that $\mathbf{x} = \min(\mathbf{u}, \mathbf{w})$ and $\mathbf{y} = \min(\mathbf{v}, \mathbf{w})$.

The second model to be considered is a nonfatal shock model. The shocks occur according to the same mechanism as before but now the shocks destroy the components with probabilities less than 1. A type 1 shock destroys the first component with probability p_1 and a type 2 shock destroys the second component with probability p_2. Type 3 shocks destroy components according to the following table:

Components Destroyed	*Probability*
Both	p_{00}
First only	p_{01}
Second only	p_{10}
Neither	p_{11}

We have for $t \geq s \geq 0$,

$$\begin{aligned}P(\mathbf{x} > s, \mathbf{y} > t) &= \left\{\sum_{k=0}^{\infty} e^{-\lambda_1 s} \frac{(\lambda_1 s)^k}{k!} (1 - p_1)^k\right\}\left\{\sum_{\ell=0}^{\infty} e^{-\lambda_2 t} \frac{(\lambda_2 t)^\ell}{\ell!} (1 - p_2)^\ell\right\} \\ &\quad \cdot\left\{\sum_{m,n=0} \left[e^{-\lambda_{12} s} \frac{(\lambda_{12} s)^m}{m!} p_{11}{}^m\right]\right. \\ &\qquad\qquad \left.\cdot\left[e^{-\lambda_{12}(t-s)} \frac{(\lambda_{12}(t-s))^n}{n!} (p_{11} + p_{01})^n\right]\right\} \\ &= \exp\left[-\lambda_1 s + \lambda_1 s(1 - p_1) - \lambda_2 t + \lambda_2 t(1 - p_2)\right. \\ &\qquad \left. - \lambda_{12} t + \lambda_{12} s p_{11} + \lambda_{12}(t - s)(p_{11} + p_{01})\right] \\ &= \exp(-s\lambda_1{}^* - t\lambda_2{}^* - t\lambda_{12}{}^*)\end{aligned}$$

where $\lambda_1^* = \lambda_1 p_1 + \lambda_{12} p_{01}$, $\lambda_2^* = \lambda_2 p_2 + \lambda_{12} p_{10}$, and $\lambda_{12}^* = \lambda_{12} p_{00}$. A similar result for $s \geq t \geq 0$ yields (9.3) with λ_1, λ_2, and λ_{12} replaced by λ_1^*, λ_2^*, and λ_{12}^*.

The third and final derivation of the BVE generalizes the "no deterioration" property of the univariate exponential. The generalization which proves useful in two and more dimensions is

$$P(\mathbf{x} > s_1 + t, \mathbf{y} > s_2 + t \mid \mathbf{x} > s_1, \mathbf{y} > s_2) = P(\mathbf{x} > t, \mathbf{y} > t) \quad (9.4)$$

or, in words, if two components have ages exceeding s_1 and s_2 then the probability that both components are functioning t time units later is the same as if both components were new. Equation (9.4) is equivalent to

$$\bar{F}(s_1 + t, s_2 + t) = \bar{F}(s_1, s_2) \cdot \bar{F}(t, t). \quad (9.5)$$

Theorem 9.6

x and **y** are BVE if they have exponential marginals and their joint survival function satisfies (9.5).

Proof

Setting $s_1 = s_2 = s$ in (9.5) yields $\bar{F}(s + t, s + t) = \bar{F}(s, s) \cdot \bar{F}(t, t)$ which implies $\bar{F}(s, s) = \exp(-\theta s)$ for some $\theta > 0$. Thus, $\bar{F}(s + t, t) = \bar{F}(s, 0) \cdot e^{-\theta t}$. The requirement of exponential marginals yields

$$\bar{F}(s + t, t) = \exp[-(\delta_1 s + \theta t)]$$

or

$$\bar{F}(x, y) = \exp[-\theta y - \delta_1(x - y)] \quad \text{for } x \geq y.$$

Similarly for $x \leq y$ we obtain $\bar{F}(x, y) = \exp[-\theta x - \delta_2(y - x)]$. Substituting $\theta = \lambda_1 + \lambda_2 + \lambda_{12}$, $\delta_1 = \lambda_1 + \lambda_{12}$ and $\delta_2 = \lambda_2 + \lambda_{12}$ gives

$$\bar{F}(x, y) = \exp[-(\lambda_1 x + \lambda_2 y + \lambda_{12} \max(x, y))], \quad x, y \geq 0.$$

Monotonicity of $\bar{F}$ implies $\lambda_1, \lambda_2 \geq 0$ but it is apparently not so easy to prove $\lambda_{12} \geq 0$. We refer the reader to Marshall and Olkin's original paper for the proof of this final point.

The marginal survival functions are easily computed to be

$$\bar{F}_{\mathbf{x}}(s) = \bar{F}(s, 0) = \exp[-(\lambda_1 + \lambda_{12})s], \quad s \geq 0$$

and

$$\bar{F}_{\mathbf{y}}(t) = \bar{F}(0, t) = \exp[-(\lambda_2 + \lambda_{12})t], \quad t \geq 0.$$

To obtain the moments we compute

$$E\mathbf{x} = (\lambda_1 + \lambda_{12})^{-1}, \quad V(\mathbf{x}) = (\lambda_1 + \lambda_{12})^{-2}$$
$$E\mathbf{y} = (\lambda_2 + \lambda_{12})^{-1}, \quad V(\mathbf{y}) = (\lambda_2 + \lambda_{12})^{-2}$$

from the marginal distributions and by a lengthy direct integration

$$E\mathbf{xy} = \frac{1}{\lambda}\left(\frac{1}{\lambda_1 + \lambda_{12}} + \frac{1}{\lambda_2 + \lambda_{12}}\right)$$

where $\lambda = \lambda_1 + \lambda_2 + \lambda_{12}$. The covariance and the correlation are respectively

$$C(\mathbf{x}, \mathbf{y}) = \frac{\lambda_{12}}{\lambda(\lambda_1 + \lambda_{12})(\lambda_2 + \lambda_{12})}$$

and $\rho = \lambda_{12}/\lambda$. Note that $0 \leq \rho \leq 1$ and that $\mathbf{x}$ and $\mathbf{y}$ are uncorrelated if and only if they are independent.

Two conditional probabilities can be calculated. We have

$$P(\mathbf{x} > s \,|\, \mathbf{y} = t) = \lim_{\Delta \to 0} \frac{\bar{F}(s, t) - \bar{F}(s, t + \Delta)}{\bar{F}(0, t) - \bar{F}(0, t + \Delta)}$$

$$= \begin{cases} e^{-\lambda_1 s}, & s \leq t \\ \dfrac{\lambda_2}{\lambda_2 + \lambda_{12}} e^{-\lambda_1 s - \lambda_{12}(s - t)}, & s > t \end{cases}$$

and

$$P(\mathbf{x} > s \,|\, \mathbf{y} > t) = \frac{\bar{F}(s, t)}{\bar{F}(0, t)}$$

$$= \exp\left[-\lambda_1 s - \lambda_{12} \max(s - t, 0)\right], \quad s, t > 0.$$

An important property of independent exponential r.v.'s $\mathbf{x}$ and $\mathbf{y}$ is that min $(\mathbf{x}, \mathbf{y})$ is exponential. This property continues to hold for bivariate exponential r.v.'s, since

$$P(\min(\mathbf{x}, \mathbf{y}) > s) = \bar{F}(s, s) = e^{-\lambda s},$$

$\lambda = \lambda_1 + \lambda_2 + \lambda_{12}$.

Finally we wish to note that the BVE distribution does not admit a density function, as it places positive probability on a set of zero area:

Theorem 9.7

If $\mathbf{x}$ and $\mathbf{y}$ are BVE then $P(\mathbf{x} > \mathbf{y}) = \lambda_2/\lambda$, $P(\mathbf{y} > \mathbf{x}) = \lambda_1/\lambda$, and $P(\mathbf{x} = \mathbf{y}) = \lambda_{12}/\lambda$.

Proof

For $\mathbf{x}$ and $\mathbf{y}$ independent

$$P(\mathbf{x} > \mathbf{y}) = \int_0^\infty e^{-\lambda_1 s} \cdot \lambda_2 e^{-\lambda_2 s}\, ds = \frac{\lambda_2}{\lambda_1 + \lambda_2}.$$

If $\lambda_{12} > 0$, then from the fatal shock model, $\mathbf{x} = \min(\mathbf{u}, \mathbf{w})$ and $\mathbf{y} = \min(\mathbf{v}, \mathbf{w})$ where $\mathbf{u}$, $\mathbf{v}$, and $\mathbf{w}$ are independent exponentials. Hence, $\min(\mathbf{u}, \mathbf{w})$ is exponential with parameter $\lambda_1 + \lambda_{12}$. Now

$$P(\mathbf{x} > \mathbf{y}) = P(\min(\mathbf{u}, \mathbf{w}) > \mathbf{v}) = \frac{\lambda_2}{\lambda},$$

according to the first part of this proof. The remaining probabilities are obtained by permuting indices and by subtraction.

The previous results can all be generalized to p dimensions but the notation becomes formidable; to reduce this difficulty to a minimum we adopt another method of indexing the λ's. Let V denote the set of vectors $v = (v_1, \ldots, v_p)$ where each component is 0 or 1 but $(v_1, \ldots, v_p) \neq (0, \ldots, 0)$. We define the *multivariate exponential* (MVE) survival function by the equation

$$\bar{F}(x_1, \ldots, x_p) = \exp\left[-\sum_V \lambda_{v_1 \cdots v_p} \max(x_1 v_1, \ldots, x_p v_p)\right],$$

$x_i \geq 0$. For $p = 2$ this becomes

$$\bar{F}(x_1, x_2) = \exp[-\lambda_{10}x_1 - \lambda_{01}x_2 - \lambda_{11}\max(x_1, x_2)]$$

which is the BVE previously discussed except for the labeling of the λ's.

The MVE can be obtained from the requirement of no deterioration coupled with marginal exponential distributions as well as from fatal and nonfatal shock models.

Theorem 9.8

If $\mathbf{x}_1, \ldots, \mathbf{x}_p$ are MVE, there exist independent exponential r.v.'s $\mathbf{z}_v$, $v \in V$, such that $\mathbf{x}_i = \min\{\mathbf{z}_v | v_i = 1\}$.

We may use this result to generalize Theorem 9.7 to the multivarate case.

Theorem 9.9

If $\mathbf{x}_1, \ldots, \mathbf{x}_p$ are MVE then

$$P(\mathbf{x}_1 > \mathbf{x}_2) = k \sum_{v_3, \ldots, v_p} \lambda_{01v_3 \cdots v_p}$$

$$P(\mathbf{x}_2 > \mathbf{x}_1) = k \sum_{v_3, \ldots, v_p} \lambda_{10v_3 \cdots v_p}$$

and

$$P(\mathbf{x}_2 = \mathbf{x}_1) = k \sum_{v_3, \ldots, v_p} \lambda_{11v_3 \cdots v_p}$$

where k is determined by the requirement that these three probabilities sum to 1.

Proof

We need prove only the first relation. From the above theorem

$$\mathbf{x}_1 = \min\{\mathbf{z}_v: v_1 = 1\} = \min[\min\{\mathbf{z}_v: v_1 = 1, v_2 = 0\}, \min\{\mathbf{z}_v: v_1 = 1, v_2 = 1\}]$$
$$\mathbf{x}_2 = \min\{\mathbf{z}_v: v_2 = 1\} = \min[\min\{\mathbf{z}_v: v_1 = 0, v_2 = 1\}, \min\{\mathbf{z}_v: v_1 = 1, v_2 = 1\}].$$

Thus $\mathbf{x}_1 > \mathbf{x}_2$ if and only if $\min\{\mathbf{z}_v: v_1 = 1\} > \min\{\mathbf{z}_v: v_1 = 0, v_2 = 1\}$, but these last two variables are independent exponentials with parameters

$$\lambda_1 = \sum_{v_2, \ldots, v_p} \lambda_{1v_2 \cdots v_p} \quad \text{and} \quad \lambda_2 = \sum_{v_3, \ldots, v_p} \lambda_{01v_3 \cdots v_p}.$$

Thus, $P(\mathbf{x}_1 > \mathbf{x}_2) = \lambda_2/(\lambda_1 + \lambda_2)$.

Marginal distributions are given by Theorem 9.10.

Theorem 9.10

If $\mathbf{x}_1, \ldots, \mathbf{x}_m, \mathbf{x}_{m+1}, \ldots, \mathbf{x}_p$ are MVE with parameters $\lambda_{v_1, \ldots, v_p}$, then $\mathbf{x}_1, \ldots, \mathbf{x}_m$ are MVE with parameters

$$\lambda'_{v_1 \cdots v_m} = \sum_{v_{m+1}} \cdots \sum_{v_p} \lambda_{v_1 \cdots v_m v_{m+1} \cdots v_p}.$$

Proof

For $x_i \geq 0$,

$$\bar{F}(x_1, \ldots, x_m, 0, \ldots, 0) = \exp\left[-\sum_V \lambda_{v_1 \cdots v_p} \max(x_1 v_1, \ldots, x_m v_m)\right]$$
$$= \exp\left[-\sum_{v_1} \cdots \sum_{v_m} \lambda'_{v_1 \cdots v_m} \max(x_1 v_1, \ldots, x_m v_m)\right]$$

To calculate conditional probabilities for the MVE consider

$$P = P(\mathbf{x}_1 > x_1, \ldots, \mathbf{x}_m > x_m \mid \mathbf{x}_{m+1} > x_{m+1}, \ldots, \mathbf{x}_p > x_p)$$
$$= \exp\left\{-\sum_V \lambda_v[\max(x_1 v_1, \ldots, x_p v_p) - \max(x_{m+1} v_{m+1}, \ldots, x_p v_p)]\right\}.$$

We may take $x_1 \geq \cdots \geq x_m$ and $x_{m+1} \geq \cdots \geq x_p$ without loss of generality. Let $T_1 = \{v: v_1 = \cdots = v_m = 0\}$, $T_2 = \{v: v_{m+1} = \cdots = v_p = 0\}$ and $R = V - T_1 - T_2$. Write $f(v)$ for the subscript of the first nonzero term of $(v_1, \ldots, v_m)$ and $g(v)$ for the subscript of the first nonzero term of $(v_{m+1}, \ldots, v_p)$. The following table evaluates the terms in the exponent of P.

	(a) $\max(x_1v_1, \ldots, x_pv_p)$	(b) $\max(x_{m+1}v_{m+1}, \ldots, x_pv_p)$	(a) − (b)
T_1	$x_{g(v)}$	$x_{g(v)}$	0
T_2	$x_{f(v)}$	0	$x_{f(v)}$
R	$\max(x_{f(v)}, x_{g(v)})$	$x_{g(v)}$	$\max(x_{f(v)} - x_{g(v)}, 0)$

We find

$$P = \exp\left[-\sum_{T_2} \lambda_v x_{f(v)} - \sum_{R} \lambda_v \max(x_{f(v)} - x_{g(v)}, 0)\right]$$

For example:

$$P(\mathbf{x}_1 > x_1 | \mathbf{x}_2 > x_2, \mathbf{x}_3 > x_3)$$
$$= \exp[-\lambda_{100}x_1 - (\lambda_{110} + \lambda_{111}) \max(x_1 - x_2, 0) - \lambda_{101} \max(x_1 - x_3, 0)].$$

PROBLEMS

1. Give a careful discussion of independence, being sure to cover the following points. Define independence of events, probability experiments, and random variables; show the relations between these concepts. If A and B are independent then what other events are also independent? Discuss mutual and pairwise independence giving an example if you can. Give various alternative tests to determine when random variables are independent.

2. For what values of a and b can the following be a variance-covariance matrix?

$$\begin{pmatrix} a & b & b \\ b & a & b \\ b & b & a \end{pmatrix}$$

3. Prove: If $m \to \infty$ and $n \to \infty$ so that $c_n^2/a_m^2 \to \lambda^2$ and if the bivariate distribution of the random variables $(f_m - b_m)/a_m$ and $(g_n - d_n)/c_n$ approaches that of f^* and g^* then

$$P\left(\frac{f_m - g_n - (b_m - d_n)}{(a_m^2 + c_n^2)^{1/2}} \leq z\right) \to P(f^* - \lambda g^* \leq z(1 + \lambda^2)^{1/2})$$

4. (Firestone and Hanson). Let $a = (\frac{1}{4}\pi)e^{-4}$ and

$$f(x, y) = \begin{cases} a, & \text{if } 1 < |x| < 2 \text{ and } 1 < |y| < 2, \text{ or if} \\ & \quad 0 < |x| < 1 \text{ and } 0 < |y| < 1 \\ -a, & \text{if } 1 < |x| < 2 \text{ and } 0 < |y| < 1, \text{ or if} \\ & \quad 0 < |x| < 1 \text{ and } 0 < |y| < 2 \\ 0, & \text{otherwise.} \end{cases}$$

Show that $p(x, y) = f(x, y) + (\frac{1}{2}\pi) \exp(-\frac{1}{2}(x^2 + y^2))$ is a density. Hence, r.v.'s which are individually normal need not be jointly normal. Also, uncorrelated r.v.'s with normal marginals need not be independent.

5. The joint density of the r.v.'s $\mathbf{x}$, $\mathbf{y}$, $\mathbf{z}$ is $f(x, y, z)$. Find the conditional probability density of these three r.v.'s given that $\mathbf{x} + \mathbf{y} \geq \mathbf{z}$, expressing the result as an integral.

6. Let $\mathbf{x}_1$ and $\mathbf{x}_2$ be independent and $N(\mu, \sigma^2)$. Obtain the distribution of $\mathbf{v} = (\mathbf{x}_1 + \mathbf{x}_2 - 2\mu)/(\mathbf{x}_1 - \mathbf{x}_2)$.

7. In a problem of target analysis the miss distance $\mathbf{r}$ has the radial normal density

$$f(r) = \left(\frac{r}{\sigma^2}\right) \exp\left(\frac{-r^2}{2\sigma^2}\right), \quad 0 < r < \infty$$

and $p(r)$ is the conditional probability of an event E given miss distance r. The unconditional probability of E is

$$P = \int_0^\infty p(r) f(r)\, dr.$$

Show that if $p(r)$ is of the form

$$p(r) = k \exp\left(\frac{-r^2}{2c^2}\right),$$

then $P = kc^2/(c^2 + \sigma^2)$; however, if there is a fixed aiming bias h in any direction then P is replaced by

$$P_h = \frac{kc^2}{(c^2 + \sigma^2)} \exp \frac{-h^2}{2(c^2 + \sigma^2)}.$$

8. By the method of characteristic functions prove the following result. If $\mathbf{y}_1, \ldots, \mathbf{y}_p$ have the density function $R(y_1, \ldots, y_p) \exp[-\frac{1}{2}H(y_1, \ldots, y_p)]$ where $R(y_1, \ldots, y_p) = R(y)$ and $H(y_1, \ldots, y_p) = H(y)$ are homogeneous functions of order m and q respectively, then $H(y)$ has the chi-square distribution with $2(m + p)/q$ degrees of freedom.

9. Prove: If $R(y)$ and $H(y)$ are nonnegative homogeneous functions of order m and q, respectively, then

$$\int \cdots \int_{H(y) \leq t} R(y) \exp[-\tfrac{1}{2}H(y)]\, dy_1 \cdots dy_p$$

$$= P(\chi^2_{2(m+p)/q} \leq t) \int_{-\infty}^{\infty} \cdots \int R(y) \exp[-\tfrac{1}{2}H(y)]\, dy_1 \cdots dy_p$$

whenever the integrals exist.

10. Show

$$\int \cdots \int_{\sum_i y_i^2 \leq t} \left(\sum c_i y_i^2\right)^j \exp\left(-\frac{1}{2}\sum_i y_i^2\right) dy_1 \cdots dy_p$$

$$= P(\chi^2_{p+2j} \leq t) \int_{-\infty}^{\infty} \cdots \int \left(\sum_i c_i y_i^2\right)^j \exp\left(-\frac{1}{2}\sum_i y_i^2\right) dy_1 \cdots dy_p$$

where the c's may have arbitrary sign.

11. Obtain the following expansion for the expression (9.1):
$P(\mathbf{Q} \leq t)$

$$= \sum_{j=0}^{k-1} \frac{D}{j!} \int \cdots \int_R \left[-\frac{1}{2}\sum (r_i - 1) y_i^2\right]^j \exp\left(-\frac{1}{2}\sum y_i^2\right) dy_1 \cdots dy_p + E_k.$$

R is the region $\sum y_i^2 \leq t/\mu$, $r_i = \mu/a_i$ and $\mu > 0$ is arbitrary.

12. (continuation) Show

$$P(\mathbf{Q} \leq t) = \left(\prod_i r_i\right)^{1/2} \sum_{j=0}^{k-1} \frac{E\left[-\frac{1}{2}\sum_i (r_i - 1)\mathbf{y}_i^2\right]^j}{j!} P\left(\chi^2_{p+2j} \leq \frac{t}{\mu}\right) + E_k.$$

where it is understood that the expectation is taken with respect to p independent $N(0, 1)$ deviates $\mathbf{y}_1, \ldots, \mathbf{y}_p$.

APPENDIX—VECTORS, MATRICES, AND DETERMINANTS

The purpose of this appendix is to provide ready definitions and brief discussion of concepts needed for our treatment of the MVN distribution. This material can be considered as an outline which an instructor may follow or as a review for a sophisticated audience. If there exists an independent reader who is unfamiliar with these ideas but has gotten this far, he should scan a text on matrix theory proper.[4]

Rectangular arrays of numbers are called *matrices*. For example,

$$\begin{pmatrix} 1 & -1 & 2 & 1 \\ 3 & 2 & 0 & 1 \\ 4 & 1 & 2 & 2 \end{pmatrix}$$

is a 3×4 matrix. Two *matrices* are *equal* if their corresponding elements are equal. Matrices having one row and those having one column are called *row* and *column vectors*, respectively. Vectors $A = (a_1, a_2, \ldots, a_n)$ and $B = (b_1, b_2, \ldots, b_n)$ are called *orthogonal* if $a_1b_1 + a_2b_2 + \cdots + a_nb_n = 0$. The square of the length of the vector A is $\sum_{i=1}^n a_i^2$.

[4] Mirsky, L., *An Introduction to Linear Algebra*. Oxford—at the Clarendon Press, 1955.

If c is a scalar (that is, a number) and A and B are $m \times n$ matrices then $cA = Ac$ is the matrix obtained from A by multiplying each element of A by c, and $A + B$ is the $m \times n$ matrix obtained by adding the corresponding elements of A and B. Thus

$$3\begin{pmatrix} 1 & 3 & -1 \\ 0 & \frac{1}{3} & 0 \end{pmatrix} = \begin{pmatrix} 3 & 9 & -3 \\ 0 & 1 & 0 \end{pmatrix}$$

and

$$\begin{pmatrix} 1 & 3 & -1 \\ 0 & \frac{1}{3} & 0 \end{pmatrix} + \begin{pmatrix} 2 & 4 & 1 \\ 7 & 0 & 3 \end{pmatrix} = \begin{pmatrix} 3 & 7 & 0 \\ 7 & \frac{1}{3} & 3 \end{pmatrix}.$$

Addition of matrices is defined only if the two matrices have the same number of rows and the same number of columns.

The matrix C, whose element in the ith row and jth column is c_{ij}, is frequently denoted by (c_{ij}).

If

$$A = (a_{ij}) = \begin{pmatrix} a_{11} & a_{12} & \cdots & a_{1n} \\ a_{21} & a_{22} & \cdots & a_{2n} \\ \vdots & \vdots & & \\ a_{m1} & a_{m2} & \cdots & a_{mn} \end{pmatrix}$$

and

$$B = (b_{ij}) = \begin{pmatrix} b_{11} & b_{12} & \cdots & b_{1q} \\ b_{21} & b_{22} & \cdots & b_{2q} \\ \vdots & \vdots & & \vdots \\ b_{p1} & b_{p2} & & b_{pq} \end{pmatrix}$$

then $B \cdot A$ is defined only if $q = m$. In this case $B \cdot A = C = (c_{ij})$ where

$$c_{ij} = \sum_{k=1}^{m} b_{ik} a_{kj}.$$

Thus

$$\begin{pmatrix} 1 & 3 & -1 \\ 0 & \frac{1}{3} & 0 \end{pmatrix} \cdot \begin{pmatrix} 0 & 4 & 1 \\ -\frac{1}{3} & 3 & 0 \\ -1 & 0 & 0 \end{pmatrix} = \begin{pmatrix} 0 & 13 & 1 \\ -\frac{1}{9} & 1 & 0 \end{pmatrix}$$

and

$$\begin{pmatrix} 1 & 3 \\ 0 & -1 \end{pmatrix} \begin{pmatrix} 0 \\ 1 \end{pmatrix} = \begin{pmatrix} 3 \\ -1 \end{pmatrix}$$

while

$$\begin{pmatrix} 1 \\ 3 \end{pmatrix} \cdot \begin{pmatrix} 1 & 3 \\ 0 & -1 \end{pmatrix}$$

is not defined. Roughly speaking, multiplication is carried out row by column. As in multiplying ordinary numbers, we sometimes find it convenient to omit the multiplication sign, writing AB instead of $A \cdot B$.

A matrix of the form

$$\begin{pmatrix} d_1 & 0 & \cdots & 0 \\ 0 & d_2 & \cdots & 0 \\ \vdots & \vdots & & \vdots \\ 0 & 0 & \cdots & d_n \end{pmatrix}$$

is denoted by diag $(d_1, d_2, \ldots, d_n)$ and is called a *diagonal matrix.*

m variables $y_1, y_2, \ldots, y_m$ may be expressed in terms of the n variables $x_1, x_2, \ldots, x_n$ by the relation

$$\begin{aligned} y_1 &= a_{11}x_1 + a_{12}x_2 + \cdots + a_{1n}x_n \\ y_2 &= a_{21}x_1 + a_{22}x_2 + \cdots + a_{2n}x_n \\ \vdots &\qquad \vdots \qquad\quad \vdots \qquad\qquad\quad \vdots \\ y_m &= a_{m1}x_1 + a_{m2}x_2 + \cdots + a_{mn}x_n. \end{aligned} \tag{9.6}$$

If $(x_1, x_2, \ldots, x_n)$ is interpreted as a point of $\mathbf{E}_n$, Euclidean n-space, and $(y_1, y_2, \ldots, y_m)$ is a point of $\mathbf{E}_m$, then the above equations are said to be a linear transformation of $\mathbf{E}_n$ into $\mathbf{E}_m$. If $n = m$ and the rows of the matrix

$$A = \begin{bmatrix} a_{11} & a_{12} & \cdots & a_{1n} \\ a_{21} & a_{22} & \cdots & a_{2n} \\ \vdots & & & \vdots \\ a_{n1} & a_{n2} & \cdots & a_{nn} \end{bmatrix}$$

are orthogonal vectors of length one then the matrix and corresponding linear transformation are called *orthogonal.* Geometrically an orthogonal transformation is a rotation of the coordinate system without translation or change of scale.

Equation (9.6) may be written briefly as $Y = AX$ where $A = (a_{ij})$, and X and Y are the column vectors with elements $x_1, x_2, \cdots, x_n$ and $y_1, y_2, \cdots, y_m$, respectively.

Consider the consequences of performing the transformations

$$\begin{aligned} y_1 &= a_{11}x_1 + a_{12}x_2 + a_{13}x_3 \\ y_2 &= a_{21}x_1 + a_{22}x_2 + a_{23}x_3 \end{aligned} \tag{9.7}$$

and

$$\begin{aligned} z_1 &= b_{11}y_1 + b_{12}y_2 \\ z_2 &= b_{21}y_1 + b_{22}y_2. \end{aligned} \tag{9.8}$$

By substituting (9.7) in (9.8) and collecting the coefficients of x_1, x_2, and x_3 the resulting transformation is seen to be

$$z_1 = (b_{11}a_{11} + b_{12}a_{21})x_1 + (b_{11}a_{12} + b_{12}a_{22})x_2 + (b_{11}a_{13} + b_{12}a_{23})x_3$$
$$z_2 = (b_{21}a_{11} + b_{22}a_{21})x_1 + (b_{21}a_{12} + b_{22}a_{22})x_2 + (b_{21}a_{13} + b_{22}a_{23})x_3.$$

This work can be briefly carried out as follows: $Y = AX$, $Z = BY$, $Z = BAX$, thus showing a motivation for the earlier definition of multiplication of matrices.

We now turn to a discussion of *determinants*. Any ordering $i_1, i_2, \ldots, i_n$ of the first n positive integers may be returned to the order $1, 2, \ldots, n$ by a sequence of interchanges of only two numbers. Such an ordering, called a permutation, is even or odd according as an even or odd number of such interchanges are required to convert to the normal order. $|A|$ the determinant of an $n \times n$ matrix, A, is a scalar valued function of the elements of A. $|A|$ is obtained by forming all products $a_{1i_1}a_{2i_2} \cdots a_{ni_n}$ which have one element from each row and one from each column, prefixing a $+1$ or -1 according as the permutation $i_1, i_2, \ldots, i_n$ is even or odd, and adding over all $n!$ terms. Thus,

$$|A| = \sum \pm a_{1i_1}a_{2i_2} \cdots a_{ni_n}$$

where the sign is determined by the above rule and the summation is over all $n!$ possible orderings of the numbers $1, 2, \ldots, n$.

If A and B are two square matrices, then the determinant of the product is $|A \cdot B| = |A| \cdot |B|$. This result is proved by successive application of the following three properties (which are themselves substantial results requiring some proof).

(i) If the elements of any row of A are multiplied by a constant c then the determinant of the resulting matrix is $c|A|$.

(ii) If two rows of a matrix are interchanged then the determinant changes sign.

(iii)
$$\begin{vmatrix} a_{11} & a_{12} & \cdots & a_{1n} \\ \vdots & \vdots & & \vdots \\ a_{k1} + b_{k1} & a_{k2} + b_{k2} & \cdots & a_{kn} + b_{kn} \\ \vdots & \vdots & & \vdots \\ a_{n1} & a_{n2} & \cdots & a_{nn} \end{vmatrix} = \begin{vmatrix} a_{11} & a_{12} & \cdots & a_{1n} \\ \vdots & \vdots & & \vdots \\ a_{k1} & a_{k2} & \cdots & a_{kn} \\ \vdots & \vdots & & \vdots \\ a_{n1} & a_{n2} & \cdots & a_{nn} \end{vmatrix} + \begin{vmatrix} a_{11} & a_{12} & \cdots & a_{1n} \\ \vdots & \vdots & & \vdots \\ b_{k1} & b_{k2} & \cdots & b_{kn} \\ \vdots & \vdots & & \vdots \\ a_{n1} & a_{n2} & \cdots & a_{nn} \end{vmatrix}.$$

A^T, called A *transpose*, is obtained from A by interchanging rows and columns; that is, if a_{ij} is the element in the ith row and jth column of A, then the element in the ith row and jth column of A^T is a_{ji}. If $A^T = A$ then A is said to be a *symmetric matrix*.

Several properties of the transpose will be needed. We have by direct calculation that $(AB)^T = B^T A^T$ and $|A^T| = |A|$. Also, any quadratic form

$$\sum_{i,j} b_{ij} x_i x_j$$

can be written in matrix notation as $X^T A X$ where A is a symmetric matrix. In fact, $a_{ij} = (b_{ij} + b_{ji})/2$.

An $n \times n$ matrix is said to be nonsingular if its determinant is nonzero. There are a number of important results concerning the multiplication of nonsingular matrices. Most of these are given in the following theorem.

Theorem

Let **M** be the collection of all $n \times n$ nonsingular matrices. If A, B, and C are in **M**, then

(a) AB is in **M**.
(b) $A(BC) = (AB)C$.
(c) There is a member, I, of **M** such that $AI = A$ for all A in **M**. Any matrix, I, having this property is called an *identity* of **M**.
(d) For each A in **M**, $AX = I$ has a solution X in **M**.[5] The solution to this equation is A^{-1}, the *inverse* of A.
(e) If $BA = CA$, then $B = C$.
(f) $IA = A$.
(g) $A^{-1} \cdot A = I$, and hence, A is an inverse of A^{-1}.
(h) If $AB = AC$, then $B = C$.
(k) The equations $AX = B$ and $YA = B$, respectively, have unique solutions X and Y in **M**.
(l) The matrix I having the property (c) is unique.
(m) The inverse of A is unique.
(n) $(AB)^{-1} = B^{-1}A^{-1}$.

Proof

(a) If $|A| \neq 0$ and $|B| \neq 0$ then $|AB| = |A| \cdot |B| \neq 0$.
(b) After evaluating the (i, j)th element of both sides the result follows from

$$\sum_k a_{ik} \sum_l b_{kl} c_{lj} = \sum_l \sum_k a_{ik} b_{kl} c_{lj}.$$

[5] Properties (a) to (d) are summarized by saying that **M** is a *group*.

(c) $I = \text{diag}\,(1, 1, \ldots, 1)$.

(d) We may, for example, treat the equation $AX = I$ as n sets of n linear equations each set to be solved by Cramer's rule.

(e) $BAA^{-1} = CAA^{-1}$; $BI = CI$; $B = C$.

(g) $A^{-1}\cdot(AA^{-1}) = A^{-1}$; $(A^{-1}\cdot A)A^{-1} = IA^{-1}$; $A^{-1}\cdot A = I$ according to (e).

(h) $A^{-1}\cdot(AB) = A^{-1}\cdot(AC)$; $B = C$.

(k) $X = A^{-1}B$ is a solution. If X^* is any other solution, then $AX = AX^*$ and $X = X^*$.

(n) $(AB)(B^{-1}\cdot A^{-1}) = AIA^{-1} = I$.

We can now discuss properties of an orthogonal transformation $Y = AX$. Directly from the definition we have $AA^T = I$, hence $A^{-1} = A^T$ and the inverse transformation is $X = A^TAX = A^TY$. Also, since the volume of a grid cell is unchanged by rotating the coordinate system, we see that the Jacobian of an orthogonal transformation is 1. A formal proof of this fact follows.

$$\begin{vmatrix} \dfrac{\partial x_1}{\partial y_1} & \cdots & \dfrac{\partial x_p}{\partial y_1} \\ \vdots & & \vdots \\ \dfrac{x_1}{\partial y_p} & \cdots & \dfrac{x_p}{\partial y_p} \end{vmatrix} = A.$$

$AA^T = I$, $1 = |AA^T| = |A|\,|A^T| = |A|^2$, and the absolute value of $|A|$ is 1.

Finally we mention the *characteristic roots* of an $n \times n$ matrix. They are the solutions of the equation $f(r) = |A - rI| = 0$. Expansion of the determinant yields

$$f(r) = (-r)^n + \left(\sum_i a_{ii}\right)(-r)^{n-1} + \cdots + |A| = 0,$$

a polynomial equation of nth degree having in general n solutions $r_1, r_2, \ldots, r_n$ which are not necessarily distinct. Note that the sum of the roots, called the *trace*, is the sum of the diagonal elements of A:

$$\sum_{i=1}^{n} r_i = \sum_{i-1}^{n} a_{ii}.$$

10

LIMITING EXTREME VALUE DISTRIBUTIONS

10.1 THE CLASS OF LIMITING DISTRIBUTIONS

There are many instances where one is interested in extreme rather than typical values of random variables. The engineer designing a sewer system or a dam is more concerned with extreme rather than average rainfall. To avoid going bankrupt an insurance company must determine its premium to withstand maximum rather than typical claims. And a statistician wishing to detect a laboratory with faulty testing procedures will scrutinize those laboratories having extreme determinations. Later, in Section 10.4, we will consider a more extensive example having to do with a model for highway traffic flow.

There is a useful result, in the spirit of the central limit theorem but applicable for extremes rather than averages, to the effect that for independent and identically distributed r.v.'s there are essentially only three asymptotic distributions. In its modern form this theorem is due to the two English statisticians R. A. Fisher and L. H. C. Tippett[1] and to the Russian probabilist B. V. Gnedenko.[2] The theory is couched in the language of the maximum but there is a simple relation, which will be explained later, between the distributions of the maximum and of the minimum.

[1] Fisher, R. A. and L. H. C. Tippett, Limiting forms of the frequency distribution of the largest or smallest member of a sample. *Proc. Camb. Phil. Soc.*, **24**, 180–190 (1928).

[2] Gnedenko, B. V., Sur la distribution limite du terme maximum d'une série aléatoire. *Ann. Math.*, **44**, 423–453 (1943).

NORTHWEST MISSOURI
STATE COLLEGE LIBRARY
MARYVILLE, MISSOURI

Theorem 10.1

Suppose that $\{\mathbf{x}_1, \mathbf{x}_2, \ldots\}$ are independent r.v.'s with a common d.f. F, and let $\mathbf{y}_n = \max(\mathbf{x}_1, \mathbf{x}_2, \ldots, \mathbf{x}_n)$. If $\mathbf{y}_n$ has any nondegenerate limiting distribution[3] then there exist sequences of constants $\{\alpha_n\}$ and $\{\beta_n\}$ with $\alpha_n > 0$, such that

$$G(y) = \lim_{n\to\infty} F^n(\alpha_n y + \beta_n) = \lim_{n\to\infty} P\left(\frac{\mathbf{y}_n - \beta_n}{\alpha_n} \leq y\right)$$

exists and is a nondegenerate d.f. where $G(y)$ is one of the following three functions:

$$G_1(y) = \exp(-e^{-y}), \qquad -\infty < y < \infty$$

$$G_2(y; \gamma) = \begin{cases} 0, & y \leq 0 \\ \exp(-y^{-\gamma}), & y > 0, \quad \gamma > 0 \end{cases}$$

$$G_3(y; \gamma) = \begin{cases} \exp(-(-y)^{\gamma}), & y < 0, \quad \gamma > 0 \\ 1, & y \geq 0. \end{cases}$$

Thus, loosely speaking, if $\mathbf{y}_n$ has any nondegenerate limiting distribution then we may write approximately for large n

$$P(\mathbf{y}_n \leq y) \simeq G_i\left(\frac{y - \beta_n}{\alpha_n}\right) \tag{10.1}$$

where G_i is one of the three functions of the theorem. This loose interpretation is entirely consistent with the usual application of other limit theorems. The initial distribution function F is said to be of *exponential*, *Cauchy*, or *limited type* according as the limit is G_1, G_2, or G_3. Of the three types the first is particularly important since it includes many of the common statistical distributions; the normal, lognormal, logistic, and exponential distributions are all of exponential type.

We now examine in turn the graphs and the moments of the three extreme value distributions.

The *double exponential distribution*

$$G_1(y) = \exp(-e^{-y}), \quad -\infty < y < \infty$$

has density function (shown in Figure 10.1)

$$g_1(y) = \exp -(y + e^{-y}), \quad -\infty < y < \infty$$

[3] $G(y)$ is a limiting distribution of $\mathbf{y}_n$ if there exist sequences of constants $\{a_n\}$ and $\{b_n\}$ with $(a_n > 0)$ such that

$$P\left(\frac{\mathbf{y}_n - b_n}{a_n} \leq y\right) \to G(y) \qquad \text{as } n \to \infty.$$

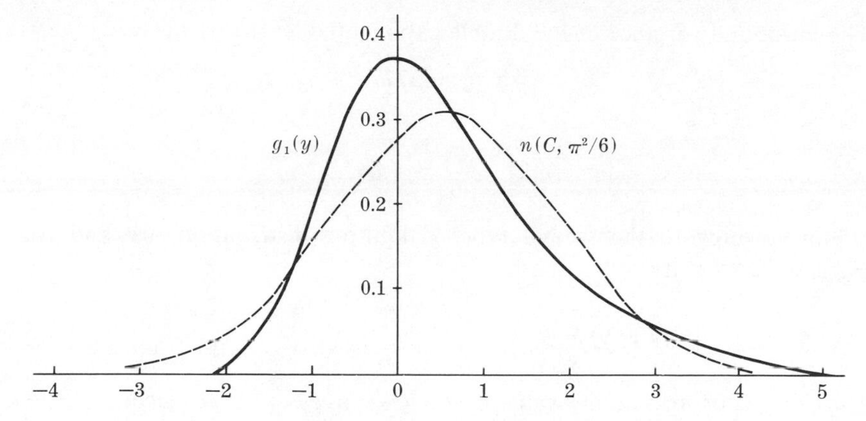

Figure 10.1 The Density Function of the Double Exponential Distribution (with $n(y; C, \pi^2/6)$ Presented for Comparison)

and c.f.

$$\phi(t) = \int_{-\infty}^{\infty} e^{ity} e^{-y} e^{-e^{-y}}\, dy.$$

Making the transformation $w = e^{-y}$, and extending the definition of the gamma function to complex arguments,

$$\begin{aligned} \phi(t) &= \int_0^{\infty} w^{-it} e^{-w}\, dw \\ &= \Gamma(1 - it) \\ &= \sum_{k=0}^{\infty} c_k (it)^k, \quad |t| < 1 \end{aligned}$$

where $c_0 = 1$, and

$$c_{n+1} = \frac{\sum_{k=0}^{n} s_{k+1} c_{n-k}}{n + 1}.$$

$s_1 = C =$ Euler's constant, and $s_n = \zeta(n)$ (the Riemann Zeta function), for $n \geq 2$.[4] We obtain $c_0 = 1$, $c_1 = C$, $c_2 = [C^2 + \zeta(2)]/2$, so that

$$\phi(t) = 1 + C \cdot it + [C^2 + \zeta(2)] \frac{(it)^2}{2} + \cdots.$$

[4] Nielsen, N., *Handbuch der Theorie der Gammafunction*, Leipzig (1906).

The mean and variance of the double exponential are respectively

$$C \simeq 0.577$$

and

$$\zeta(2) = \frac{\pi^2}{6} \simeq 1.645.$$

Now turning to the second type of limiting distribution, we find that $G_2(y, \gamma)$ has density

$$g_2(y; \gamma) = \begin{cases} 0, & y \le 0 \\ \gamma y^{-\gamma-1} e^{-y^{-\gamma}}, & y > 0, \gamma > 0. \end{cases}$$

The modal or maximum value is $\gamma[(\gamma + 1)/\gamma e]^{(\gamma+1)/\gamma}$. It occurs at $y = [\gamma/(\gamma + 1)]^{1/\gamma}$. Note that as $\gamma \to \infty$

$$G_2(y, \gamma) \to \begin{cases} 0, & y < 1 \\ e^{-1}, & y = 1 \\ 1, & y > 1 \end{cases}$$

and $g_2(y, \gamma)$ will approach an infinite spike located at $y = 1$; this tendency can be seen in Figure 10.2.

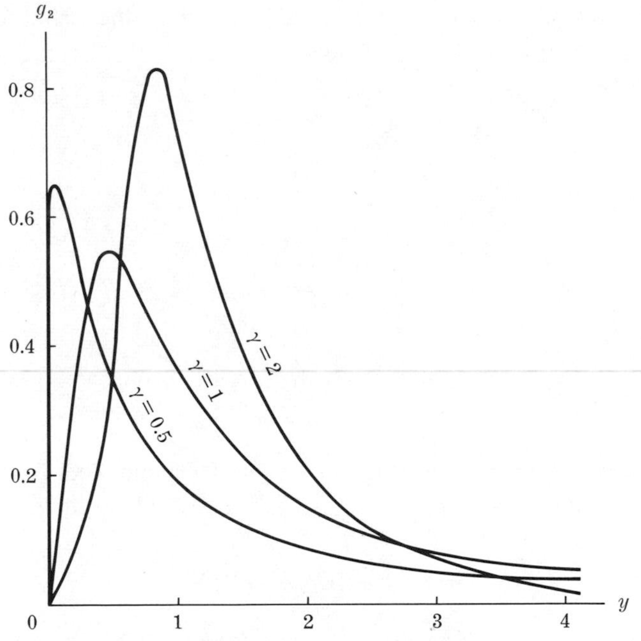

Figure 10.2 Graphs of the Density Function $g_2(y, \gamma)$

To find the kth moment write

$$\begin{aligned}\xi_k &= \int_0^\infty y^k g_2(y;\gamma)\,dy \\ &= \int_0^\infty \gamma y^{k-\gamma-1} e^{-y^{-\gamma}}\,dy \\ &= \int_0^\infty w^{-k/\gamma} e^{-w}\,dw \\ &= \Gamma\left(1-\frac{k}{\gamma}\right) \quad \text{for } k<\gamma.\end{aligned}$$

Moments of order $\geq \gamma$ do not exist.

Finally, the distribution $G_3(y, \gamma)$ has density

$$g_3(y;\gamma) = \begin{cases}\gamma(-y)^{\gamma-1}e^{-(-y)^\gamma}, & y \leq 0, \gamma > 0 \\ 0, & y > 0\end{cases}$$

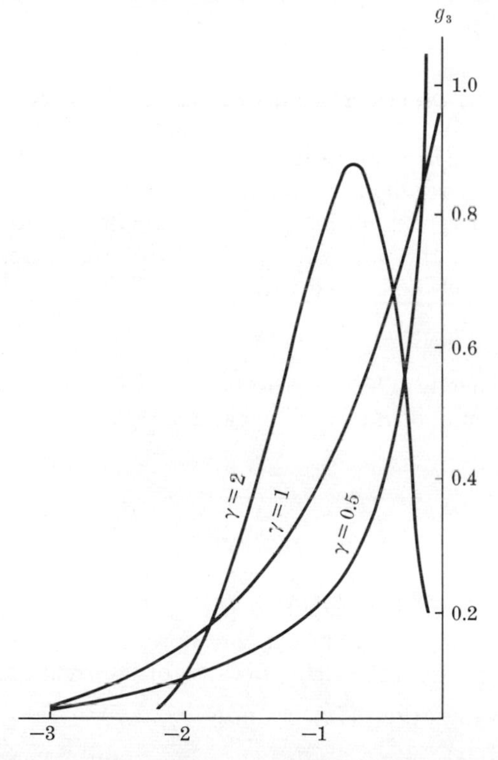

Figure 10.3 Graphs of the Density Function $g_3(y; \gamma)$

For $\gamma \geq 1$, $g_3(y; \gamma)$ assumes its maximum value $\gamma[(\gamma - 1)/\gamma e]^{(\gamma-1)/\gamma}$ at $y = -[(\gamma - 1)/\gamma]^{1/\gamma}$. For $\gamma < 1$, $g_3(y; \gamma)$ is asymptotic to the g axis and has mode 0. Note that, as $\gamma \to \infty$, $g_3(y; \gamma)$ also approaches an infinite spike, this time located at $y = -1$. (See Figure 10.3.)

The kth moment is given by

$$\xi_k = (-1)^k \Gamma\left(\frac{k}{\gamma} + 1\right) \quad \text{for } k > -\gamma.$$

For $k \leq -\gamma$ the moments do not exist.

Using the above calculations, some interpretations of the parameter $\{\alpha_n\}$ and $\{\beta_n\}$ are possible in terms of the limiting distribution (10.1). The double exponential d.f. $G_1[(y - \beta_n)/\alpha_n]$ has mode β_n while α_n, being a scale parameter, is a measure of dispersion; for example, the standard deviation is $\pi \cdot \alpha_n/\sqrt{6}$. For an initial distribution of the Cauchy type we will show, in Section 10.3, that β_n may be taken to be 0 so that the limit will have the form $G_2(y/\alpha_n; \gamma)$, which has mode $\alpha_n[\gamma/(\gamma + 1)]^{1/\gamma}$. Finally, when the limit is of the third type, we will show that $\beta_n = y^0$ (the upper bound of $\mathbf{x}_i$). The mode of this limiting distribution, $G_3[(y - y^0)/\alpha_n; \gamma]$, is y^0 for $0 < \gamma \leq 1$ and $y^0 - \alpha_n[(\gamma - 1)/\gamma]^{1/\gamma}$ for $\gamma > 1$.

10.2 PROOF OF THE MAIN THEOREM

In this section we will attempt to be somewhat more complete than usual as the details are not as readily available elsewhere. For that reason a reader primarily interested in applications may wish to skip directly to Section 10.3.

Before proceeding to the proof of Theorem 10.1 we present some preliminary results not solely concerned with extreme values.

Lemma 10.1

If F is a nondegenerate d.f. and $F(ax + b) = F(\alpha x + \beta)$ with $a, \alpha > 0$ at all points of continuity of F then $a = \alpha$ and $b = \beta$.

Proof

First we show that $F(ax + b) = F(\alpha x + \beta)$ on the entire real line. The set of points x such that either $F(ax + b)$ or $F(\alpha x + \beta)$ is discontinuous, is the union of two denumerable sets and hence is denumerable. For each point x we may choose a sequence x_n such that $\{ax_n + b\}$ and $\{\alpha x_n + \beta\}$ are continuity points of F and $x_n \to x_+$. Since $a, \alpha > 0$, we have $ax_n + b \to ax + b$ and $\alpha x_n + \beta \to \alpha x + \beta$ from the right. Because of the right continuity of F

$$F(ax + b) = \lim_n F(ax_n + b) = \lim_n F(\alpha x_n + \beta) = F(\alpha x + \beta)$$

which achieves our first objective.

From the hypothesized functional relation we obtain $F(x) = F(Ax + B)$ where $A = a/\alpha > 0$ and $B = b - \beta a/\alpha$. Then for all positive and negative integers k

$$F(x) = F[A^k x + B(1 + A + \cdots + A^{k-1})].$$

If $a = \alpha$ then $A = 1$ and $F(x) = F(x + kB)$ where without loss of generality we take $B \geq 0$. If $B > 0$, then $0 = \lim_{k\to-\infty} F(x + kB) = F(x) = \lim_{k\to\infty} F(x + kB) = 1$. We must therefore have $B = 0$, that is, $b = \beta$. In the remaining case where $a \neq \alpha$, assume without loss of generality that $a < \alpha$. Then

$$F(x) = F[A^k(x - x^0) + x^0]$$

where we have written $x^0 = B/(1 - A)$. Allowing $k \to \infty$, we obtain $F(x) = F(x^0)$ for $x \geq x^0$ and $F(x) = F(x_-{}^0)$ for $x < x^0$. Hence, F is degenerate at x_0, a contradiction.

Lemma 10.2 (Khintchine)

If the sequence $\{F_n(x)\}$ of d.f.'s converges as $n \to \infty$ to a d.f. $F(x)$ then, for any choice of the constants $a_n > 0$ and b_n, the sequence $\{F_n(a_n x + b_n)\}$ converges to a nondegenerate d.f., say $G(x)$, if and only if $G(x) = F(ax + b)$ where $0 < a < \infty$, $-\infty < b < \infty$, $a_n \to a$, and $b_n \to b$.

Proof

The "if" portion of the lemma is immediate. To prove the converse let $\{a_n'\}$ and $\{b_n'\}$ be convergent subsequences of $\{a_n\}$ and $\{b_n\}$, say $a_n' \to a$ and $b_n' \to b$. For n sufficiently large we have

$$ax + b - \epsilon \leq a_n'x + b_n' \leq ax + b + \epsilon$$

so that

$$F_n(ax + b - \epsilon) \leq F_n(a_n'x + b_n') \leq F_n(ax + b + \epsilon).$$

Taking limits as $n \to \infty$

$$F(ax + b - \epsilon) \leq G(x) \leq F(ax + b + \epsilon)$$

at continuity points. Allowing $\epsilon \to 0$, $F(ax + b) = G(x)$ at continuity points. If $a = 0$ or $b = \pm\infty$, then $G(x) = F(b)$ for all x, and G cannot be a probability distribution. If $a = +\infty$ then

$$G(x) = \begin{cases} 0, & x < 0 \\ 1, & x > 0 \end{cases}$$

and G is degenerate at 0, again a contradiction.

Now suppose at least one of the sequences $\{a_n\}$ and $\{b_n\}$ does not converge. We may then choose a second pair of subsequences $\{a_n''\}$ and $\{b_n''\}$ such that $a_n'' \to \alpha$, $b_n'' \to \beta$, and either $a \neq \alpha$ or $b \neq \beta$. Applying our earlier argument to these new subsequences, we obtain $F(\alpha x + \beta) = G(x)$, and, hence $F(\alpha x + \beta) = F(ax + b)$ at points of continuity. From Lemma 10.1, we must have $a = \alpha$ and $b = \beta$ contrary to the definitions of α and β. The sequence $\{a_n\}$ and $\{b_n\}$ must therefore be convergent.

Lemma 10.3

For a sequence of d.f.'s $\{F_n(x)\}$ and a nondegenerate d.f. $F(x)$, the relations

$$F_n(a_n x + b_n) \to F(x), \quad a_n > 0$$
$$F_n(\alpha_n x + \beta_n) \to F(x), \quad \alpha_n > 0$$

are satisfied if and only if $\alpha_n/a_n \to 1$ and $(\beta_n - b_n)/a_n \to 0$.

Proof

Let $G_n(x) = F_n(a_n x + b_n)$, then $G_n(x) \to F(x)$ and $G_n[(\alpha_n/a_n)x + (\beta_n - b_n)/a_n] = F_n(\alpha_n x + \beta_n) \to F(x)$. The result now follows from a direct application of Khintchine's lemma.

To prove Theorem 10.1, Fisher and Tippett observed directly from the relation

$$F^n(a_n y + b_n) \to G(y)$$

that the limit distribution must satisfy a functional equation

$$G(y) = G^n(c_n y + d_n) \tag{10.2}$$

for all positive integers n. We prove this from Khintchine's lemma. Write $G_n(y) = F^n(a_n y + b_n)$, then $G_n(y) \to G(y)$ and

$$G_n\left(\frac{a_{ni}}{a_n} y + \frac{b_{ni} - b_n}{a_n}\right) = F^n(a_{ni} y + b_{ni}) \to G^{1/i}(y),$$

a nondegenerate distribution function. Now from Khintchine's lemma taking $n \to \infty$, $G^{1/i}(y) = G(c_i y + d_i)$ where $a_{ni}/a_n \to c_i > 0$ and $(b_{ni} - b_n)/a_n \to d_i$. This is (10.2).

The substantial remainder of the proof involves showing that only functions of the extreme value type can be solutions of the functional equation (10.2). First we note that

$$\begin{aligned} G(c_{mn} y + d_{mn}) &= G^{1/mn}(y) \\ &= G^{1/n}(c_m y + d_m) = G[c_n(c_m y + d_m) + d_n] \\ &= G[c_m c_n y + (c_n d_m + d_n)] \end{aligned}$$

and, from Lemma 10.1,

$$c_{mn} = c_m c_n$$
$$d_{mn} = c_n d_m + d_n.$$

Next observe that the class of extreme value distributions is closed under translation and change of scale.

Lemma 10.4

If G satisfies (10.1), then $H(y) = G(ay + b)$ satisfies the same equation with c_n unchanged but d_n replaced by $[d_n + b(c_n - 1)]/a$. In fact

$$\begin{aligned} H^n(y) &= G^n(ay + b) = G[c_n(ay + b) + d_n] \\ &= H\left[\frac{c_n(ay + b) + d_n - b}{a}\right] \\ &= H\left[c_n y + \frac{d_n + b(c_n - 1)}{a}\right]. \end{aligned}$$

Now if for some $n > 1$, $c_n \neq 1$, then $y_0 = c_n y_0 + d_n$, where $y_0 = d_n/(1 - c_n)$. Hence, $G(y_0) = G^n(y_0)$ and $G(y_0) = 1$ or 0; we say in this case that G is limited above or below. If $G(y_0) = 1$ (or 0) then, from Lemma 10.4,

$$\begin{aligned} H^n(y) &= G^n(y + y_0) \\ &= H[c_n y + d_n + y_0(c_n - 1)] \\ &= H(c_n y) \end{aligned}$$

and $H(0) = G(y_0) = 1$ (or 0).

That is, in the limited cases, we may assume without loss of generality that $G(0) = 1$ or 0. On the other hand, if G is unlimited, then $c_n = 1$ for all n. Thus, the solution of (10.2) falls into three cases according as G is unlimited, limited above, or limited below:

$$G^n(y) = G(y + d_n) \tag{10.3}$$

$$G(0) = 1, \quad G^n(y) = G(c_n y) \tag{10.4}$$

$$G(0) = 0, \quad G^n(y) = G(c_n y). \tag{10.5}$$

Consider the Equation (10.3), valid for all integral n, where $G(x)$ is a right continuous distribution function. First we solve the functional equation $d_{mn} = d_m + d_n$; we do this by extending d_n to the positive reals and showing that the extension must have the form $d_z = c \log z$. The restriction of d_z to the integers must then have the same form: $d_n = c \log n$.

From (10.3), for integral m and n, we have

$$G^{1/n}(y) = G(y - d_n)$$
$$G^{m/n}(y) = G^m(y - d_n) = G(y + d_m - d_n).$$

Define $d_{m/n} = d_m - d_n$. Now

$$\begin{aligned} d_{(m_1/n_1)(m_2/n_2)} &= d_{m_1 m_2} - d_{n_1 n_2} \\ &= d_{m_1} + d_{m_2} - d_{n_1} - d_{n_2} \\ &= d_{m_1/n_1} + d_{m_2/n_2}. \end{aligned}$$

Thus, $G^r(y) = G(y + d_r)$ and $d_{r_1 r_2} = d_{r_1} + d_{r_2}$ for rational r, r_1, and r_2.

Now let $r_1 < r_2 < r_3 < \cdots$ be a sequence of rationals approaching $z(> 0)$ from below. We have $G(y + d_{r_{i+1}}) = G^{r_{i+1}}(y) \leq G^{r_i}(y) = G(y + d_{r_i})$ so $d_{r_1} \geq d_{r_2} \geq \cdots$. Also, $G(y + d_{r_i}) = G^{r_i}(y) \to G^z(y)$. If $\{d_{r_i}\}$ is unbounded below then $G(y + d_{r_i}) \to 0$ and $G^z(y) = G(y) \equiv 0$. Therefore, $\{d_{r_i}\}$ is a bounded monotone sequence and approaches a limit, say d_z. Further, $G(y + d_{r_i}) \to G(y + d_z)$ so that for z_1, z_2, and z real, we have

$$G^z(x) = G(x + d_z)$$

and

$$d_{z_1 z_2} = d_{z_1} + d_{z_2}$$

where d_z is a left continuous function. This implies that d_z is of the required form $d_z = c \log z$.

Returning to Equation (10.3) we have

$$\begin{aligned} n \log G(y) &= \log G(y + d_n) \\ \log n + \log(-\log G(y)) &= \log(-\log G(y + d_n)) \end{aligned}$$

and

$$\log(-\log G(y)) - \frac{y}{c} = \log(-\log G(y + d_n)) - \frac{y + d_n}{c}$$

so that $f(y) \equiv \log(-\log G(y)) - y/c$ is periodic with period $c \log n$ for every n. But $f(y)$ is also left continuous; we show that f must therefore be constant.

Given y_1 and $\epsilon > 0$ there exists $\delta > 0$ such that if $y_1 - \delta < y < y_1$ then $|f(y) - f(y_1)| < \epsilon$ but the rationals are dense on the line so that for given y_1 and y_2 we may choose n_1 and n_2 satisfying

$$y_1 - \delta < y_2 + c \log \frac{n_1}{n_2} < y_1.$$

Now

$$\begin{aligned} f(y_2) &= f(y_2 + c \log n_1) = f(y_2 + c \log n_1 - c \log n_2) \\ &= f\left(y_2 + c \log \frac{n_1}{n_2}\right) \end{aligned}$$

and $|f(y_1) - f(y_2)| < \epsilon$ for all ϵ. Therefore, $f(y_1) = f(y_2)$. Writing $f(y) =$ constant we obtain after changing the scale

$$G(y) = \exp(-e^{-y}).$$

Equations (4.10) and (5.10) can be solved by entirely analogous methods. Fisher and Tippett show that the remaining two extreme value distributions are obtained as solutions.

10.3 THE MANNER OF THE CONVERGENCE

In Section 10.1 the parameters $\{\alpha_n\}$ and $\{\beta_n\}$ were interpreted in terms of the limiting distribution. A second and independent way of interpreting the normalizing constants involves the concept of *characteristic largest value* denoted by u_n and defined by the equation

$$n[1 - F(u_n)] = 1. \tag{10.6}$$

To avoid complications here we assume that F is continuous and monotone so that (10.6) has a unique solution. To motivate the definition let $\mathbf{M}(x)$ be the number of observations out of n equaling or exceeding x.

$$\begin{aligned} E\mathbf{M}(x) &= \sum_{i=0}^{n} i\binom{n}{i}[1 - F(x)]^i[F(x)]^{n-i} \\ &= n[1 - F(x)]. \end{aligned}$$

The expected number of observations at least as large as u_n is 1.

Next we show that the normalizing sequence $\{\alpha_n\}$ and $\{\beta_n\}$ are determined up to an asymptotic equivalence by the initial distribution. To see this we need Lemma 10.5.

Lemma 10.5 (Gnedenko)

In order that

$$F^n(\alpha_n y + \beta_n) \to G(y) \tag{10.7}$$

for all y, it is necessary and sufficient that

$$n[1 - F(\alpha_n y + \beta_n)] \to -\log G(y)$$

for all y such that $G(y) > 0$.

Proof

Equation (10.7) is equivalent to $-n \log F \to -\log G$ for $G > 0$ but from Taylor's theorem with remainder $-n \log F = n(1 - F)(1 + o(1))$.

Now if F is of *exponential type* then

$$n[1 - F(\alpha_n y + \beta_n)] \to e^{-y}, \quad -\infty < y < \infty \tag{10.8}$$

hence, $n[1 - F(\beta_n)] \to 1$ and for convergence of the first type β_n is the characteristic largest value, that is, $\beta_n = u_n$. α_n is then determined by $n[1 - F(\alpha_n + u_n)] \to e^{-1}$. To interpret α_n we proceed formally as follows. Differentiate (10.8) with respect to y to obtain

$$n\alpha_n F'(\alpha_n y + \beta_n) \to e^{-y},$$

and taking $y = 0$,

$$n\alpha_n F'(u_n) \to 1.$$

Now differentiate (10.6) with respect to n to obtain

$$-nF'(u_n)\frac{du_n}{dn} + [1 - F(u_n)] = 0$$

$$\frac{1}{nF'(u_n)} = n\frac{du_n}{dn} = \frac{du_n}{d\log n}$$

and finally

$$\alpha_n = \frac{du_n}{d\log n}(1 + o(1)).$$

Thus, α_n measures the increase of the characteristic largest value with respect to $\log n$. If α_n is independent of n, then u_n increases as $\log n$ while u_n increases faster or slower than $\log n$ according as α_n is an increasing or decreasing function of n. For the exponential type it follows that $\log n$, rather than n itself, is the scale against which to measure the increase of the largest value.

If F is of *Cauchy type* then

$$n[1 - F(\alpha_n y + \beta_n)] \to y^{-\gamma}, \quad y > 0, \gamma > 0.$$

For $c > 1$,

$$[nc][1 - F(\alpha_n y + \beta_n)] \to cy^{-\gamma}$$

where $[nc]$ is the integral part of nc. Making the change of variable $y = zc^{1/\gamma}$ one has

$$[nc][1 - F(\alpha_n c^{1/\gamma} z + \beta_n)] \to z^{-\gamma}$$

for all $z > 0$. Also

$$[nc][1 - F(\alpha_{[nc]} z + \beta_{[nc]}] \to z^{-\gamma}.$$

Lemmas 10.3 and 10.5 now yield

$$\frac{\alpha_{[nc]}}{\alpha_n c^{1/\gamma}} \to 1 \quad \text{and} \quad \frac{\beta_{[nc]}}{\alpha_{[nc]}} - \frac{\beta_n}{\alpha_n c^{1/\gamma}} \to 0.$$

Fixing n and taking $c \to \infty$ leads to

$$\frac{\beta_{[nc]}}{\alpha_{[nc]}} \to 0 \quad \text{which gives} \frac{\beta_n}{\alpha_n} \to 0.$$

Thus, relation (10.1) becomes

$$P(\mathbf{y}_n \leq y) \simeq G_2\left(\frac{y}{\alpha_n}; \gamma\right)$$

and in this sense β_n may be taken as 0.

Lemmas 10.3 and 10.5 now imply

$$n[1 - F(\alpha_n y)] \to y^{-\gamma}$$

for $y > 0$ and in particular $n[1 - F(\alpha_n)] \to 1$ so that α_n is the characteristic largest value.

If F is of *limited type* then

$$F^n(\alpha_n y + \beta_n) \to G_3(y; \gamma),$$
$$F^{2n}(\alpha_n y + \beta_n) \to G_3{}^2(y; \gamma) = G_3(2^{1/\gamma}y; \gamma),$$

and

$$F^{2n}\left(\frac{\alpha_n}{2^{1/\gamma}} y + \beta_n\right) \to G_3(y; \gamma).$$

Hence, again from Lemma 10.3,

$$\alpha_{2n} \sim \frac{\alpha_n}{2^{1/\gamma}}, \quad \beta_{2n} \sim \beta_n$$

so that

$$\alpha_{n2^k} \to 0, \quad \beta_{n2^k} \sim \beta_n. \tag{10.9}$$

Now $F^n(\beta_n) \to 1$ and if $F(y) < 1$ for all y, then $\beta_n \to \infty$; this contradicts (10.9). We have therefore proved that, for a distribution of limited type, there exists a finite number y_0 such that $F(y^0) = 1$ but $F(y^0 - \epsilon) < 1$ for each $\epsilon > 0$. Further, we may show that $\beta_n \to y^0$ and $1 - F(y^0 - \alpha_n) \sim n^{-1}$. In fact

$$n[1 - F(\alpha_n y + \beta_n)] \to \begin{cases} 0, & y \geq 0 \\ (-y)^\gamma, & y < 0 \end{cases}$$

and in particular $n[1 - F(\beta_n)] \to 0$ and $n[1 - F(\beta_n - \alpha_n)] \to 1$. The first of the above relations yields $\beta_n \geq y^0$ asymptotically while the second, using (10.9), gives $\beta_n \leq y^0$ and then $n[1 - F(y^0 - \alpha_n)] \to 1$.

If F is of limited type, y^0 being the upper limit, then $\beta_n \to y^0$ and $y^0 - \alpha_n$ is the characteristic largest value. Table 10.1 summarizes the nature of the normalizing constants of all three distribution types for the maximum.

Table 10.1 Approximate Limiting d.f. for the Maximum of n Independent r.v.'s

(u_n is the characteristic largest value.)

Initial Distribution Type	*Limit Distribution*	*Interpretation of Parameters*
exponential	$G_1\left(\frac{y - u_n}{\alpha_n}\right)$	$n[1 - F(\alpha_n + u_n)] \simeq e^{-1}, \quad \alpha_n \simeq \frac{du_n}{d \log n}$
Cauchy	$G_2(y/u_n; \gamma)$	
limited	$G_3\left(\frac{y - y^0}{y^0 - u_n}; \gamma\right)$	$y^0 = \text{l.u.b.}\,\{y \mid F(y) < 1\}$

Much more can be said about the manner of convergence to the extreme value distributions. Gnedenko has completely characterized the three types of initial distributions. We give his necessary and sufficient conditions without proof. They are:

(i) If F is continuous, then it is of exponential type, if and only if

$$\lim_{n \to \infty} n[1 - F(\alpha_n y + u_n)] = e^{-y}, \quad -\infty < y < \infty,$$

where $F(u_n) = 1 - 1/n$ and $F(\alpha_n + u_n) = 1 - 1/ne$.

(ii) F is of Cauchy type [convergence to $G_2(y, \gamma)$], if and only if,

$$\lim_{y \to \infty} \frac{1 - F(y)}{1 - F(ky)} = k^\gamma$$

for $k > 0$.

(iii) F is of limited type [convergence to $G_3(y, \gamma)$], if and only if,

$$\lim_{y \to 0-} \frac{1 - F(ky + y^0)}{1 - F(y + y^0)} = k^\gamma$$

for $k > 0$ where $F(y^0) = 1$ but $F(y^0 - \epsilon) < 1$ for all $\epsilon > 0$.

Theorem 10.1, and in fact our entire theoretical development, is stated for the maximum; the minimum may be treated using the relation

$$\min(\mathbf{x}_1, \ldots, \mathbf{x}_n) = -\max(-\mathbf{x}_1, \ldots, -\mathbf{x}_n).$$

In fact, writing $\mathbf{z}_n = \min(\mathbf{x}_1, \ldots, \mathbf{x}_n)$, then

$$P\left(\frac{\mathbf{z}_n + \beta_n}{\alpha_n} \geq x\right) = P\left(\frac{\max(-\mathbf{x}_1, \ldots, -\mathbf{x}_n) - \beta_n}{\alpha_n} \leq -x\right)$$

and if $\mathbf{z}_n$ has any nondegenerate limiting distribution

$$P(\mathbf{z}_n \le z) \simeq 1 - G_i\left(-\frac{z + \beta_n}{\alpha_n}\right) \tag{10.10}$$

will hold approximately for large n where G_i is one of the three extreme value distributions of Theorem 10.1. The density of $\mathbf{z}_n$ is $\alpha_n^{-1} g_i[(z + \beta_n)/\alpha_n]$, so that the shapes of these densities can be seen by reflecting Figures 10.1, 10.2, and 10.3 about their vertical axes.

The distribution F_- of $-\mathbf{x}_i$ determines which limit is approached and the interpretation of the normalizing constants. For continuous d.f.'s $F_-(x) = 1 - F(-x)$ and the *characteristic smallest value* 1_n is defined as the solution of

$$F(1_n) = \frac{1}{n}.$$

The characteristic smallest value of $\mathbf{x}$ is minus the characteristic largest value of $-\mathbf{x}$. Table 10.2 summarizes the results of Equation (10.10) for the three cases.

Table 10.2 Approximate Limiting d.f. for the Minimum of n Independent r.v.'s

(1_n is the characteristic smallest value.)

Type of F_-	*Limit Distribution*	*Interpretation of Parameters*
exponential	$1 - \exp\left(-e^{(z - 1_n)/\alpha_n}\right)$	$nF(1_n - \alpha_n) = e^{-1}$
Cauchy	$\begin{cases} 1, & z \ge 0 \\ 1 - \exp[-(1_n/z)^\gamma], & z < 0 \end{cases}$	
limited	$\begin{cases} 1 - \exp\left[-\left(\dfrac{z - z^0}{1_n - z^0}\right)^\gamma\right], & z > z^0 \\ 0, & z \le z^0 \end{cases}$	$z^0 = \text{g.l.b.}\ \{z \mid F(z) > 0\}$

10.4 APPLICATION—HIGHWAY TRAFFIC FLOW

In this section our intention is to construct a probabilistic model for highway traffic flow which exhibits the characteristic of congestion. That is, we want our model to have the properties of the following fundamental flow-concentration diagram of road traffic shown in Figure 10.4. Headway distribution models, like those of Section 4.9, which assume independent gaps between successive cars, cannot exhibit congestion without further assumption; there is nothing implicit in headway distribution theory to prevent the cars from going as fast as they please even though the gaps between cars are very small. We intend to examine ways in which headway distribution theory can be augmented to yield a "fundamental diagram."

Here we advance a model for the manner in which congestion affects the speed of a group of cars in close proximity all traveling in the same direction

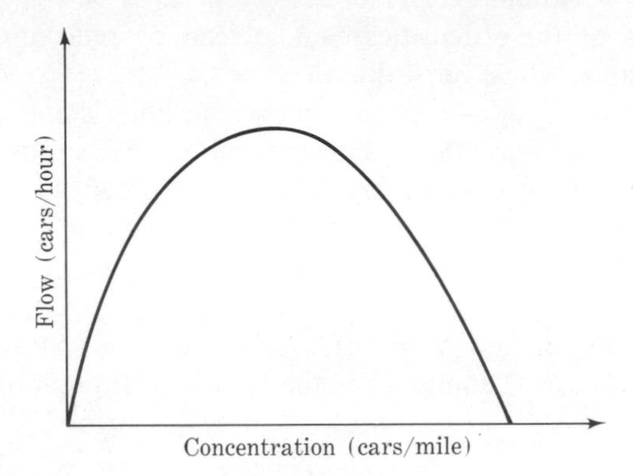

Figure 10.4

along a stretch of highway. Individual cars have desired or free speeds $v_1, v_2, \ldots$ at which they wish to travel in the absence of congestion. However, due to mutual interference or congestion, the group travels at a realized speed s_n which is the minimum of the free speeds of the group

$$s_n = \min(v_1, v_2, \ldots, v_n).$$

The model becomes probabilistic if we now consider that the free speeds are independent identically distributed r.v.'s. We find

$$P(\mathbf{s}_n \leq s) = P[\min(\mathbf{v}_1, \mathbf{v}_2, \ldots, \mathbf{v}_n) \leq s]$$

and one becomes interested in probability distributions of extreme values. We wish to approximate $P(\mathbf{s}_n \leq s)$, the distribution of realized speed.

The example we consider is that free speeds have the gamma distribution

$$\begin{aligned} F(v) &= P(\mathbf{v}_i \leq v) \\ &= \frac{\rho^\alpha}{\Gamma(\alpha)} \int_0^v r^{\alpha-1} e^{-\rho r}\, dr, \quad v > 0. \end{aligned}$$

Initially, this seems to be one of several reasonable choices. However, we will see that the resultant theory has faults which suggest that other alternatives might be considered.

Since we are working with the minimum, it is F_-, the distribution of $-\mathbf{v}_i$, which determines which limit is approached. For the gamma distribution

$$F_-(v) = 1 - F(-v) = \begin{cases} \int_{-v}^{\infty} g(r; \alpha, \rho)\, dr, & v < 0 \\ 1, & v \geq 0. \end{cases}$$

Thus, $F_-(0) = 1$ but $F_-(-\epsilon) < 1$ for $\epsilon > 0$. Also, using the mean value result

$$\int_0^x r^{\alpha-1} e^{-\rho r}\, dr = e^{-\rho \xi} \int_0^x r^{\alpha-1}\, dr, \quad 0 \leq \xi \leq x$$

we obtain

$$\lim_{x \to -0} \frac{1 - F_-(kx)}{1 - F_-(x)} = \lim_{x \to 0_+} \frac{\int_0^{kx} g(r; \alpha, \rho)\, dr}{\int_0^x g(r; \alpha, \rho)\, dr} = k^{\alpha}, \quad k > 0$$

which shows that F_- is of limited type.

Thus, we may write for the distribution of attained speed

$$P(\mathbf{s}_n \leq s) \simeq 1 - \exp\left[-\left(\frac{s}{1_n}\right)^{\alpha}\right], \quad s > 0$$

where 1_n is the characteristic smallest value of the initial distribution of free speed, that is,

$$\int_0^{1_n} g(r; \alpha, \rho)\, dr = n^{-1}.$$

We may approximate the initial distribution and estimate 1_n using a result of Fisher.[5] Fisher showed, for a χ^2 r.v. having n degrees of freedom with $n \geq 30$, that $\sqrt{2\chi^2}$ is approximately normal with mean $\sqrt{2n}$ and unit variance. For $\alpha \geq 15$ this gives

$$F(v) = \int_0^v g(r; \alpha, \rho)\, dr \simeq \int_{-\infty}^{(\sqrt{v} - \mu)/\sigma} n(t; 0, 1)\, dt$$

and

$$\int_{-\infty}^{(\sqrt{1_n} - \mu)/\sigma} n(t; 0, 1)\, dt = n^{-1}$$

where $\mu = (\alpha/\rho)^{1/2}$ and $\sigma = (4\rho)^{-1/2}$.

[5] Fisher, R. A., *Statistical Methods for Research Workers*, 8th ed., Edinburgh and London (1941).

For $\rho = \frac{1}{2}$ and $\alpha = 30$, mean free speed is 60, whereas mean realized speeds are

n	10	20	50	100	200	500	1000
l_n	46.2	42.6	39.0	36.6	34.4	32.0	25.5

Since n is the numerator of car concentration, this table shows the effect of congestion on mean realized speed according to the gamma distribution model. It appears that speed does not drop off sufficiently rapidly with concentration. We have in mind that minor congestion should reduce average realized speed but little from average free speed but that "jam packed traffic" should reduce average realized speed almost to zero.

11

STOCHASTIC PROCESSES

11.1 INTRODUCTION

A usual way of representing a p-variate distribution is by means of a Cartesian diagram as we have done for the bivariate normal at the end of Section 9.2. A second graphical means of representation is as shown in Figure 11.1. The density of $\mathbf{y}_i(s)$ is indexed by locating it at the integer i on the

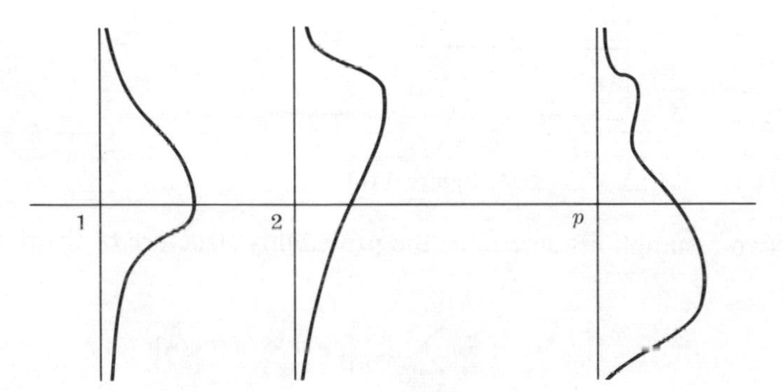

Figure 11.1

real line. It is now only a slight generalization to think of a family of r.v.'s $\{\mathbf{y}_t(s)\}$ having the entire real line as indexing set; such a family is called a *stochastic* (probability) *process*. For arbitrary fixed t, $\mathbf{y}_t(s)$ is a random variable in the sense previously discussed. For fixed $s = s_0$, $\mathbf{y}_t(s_0)$ is the result of a probability experiment; we call it a *realization* of the process. $\mathbf{y}_t(s_0)$ is an ordinary function of t as shown in Figure 11.2.

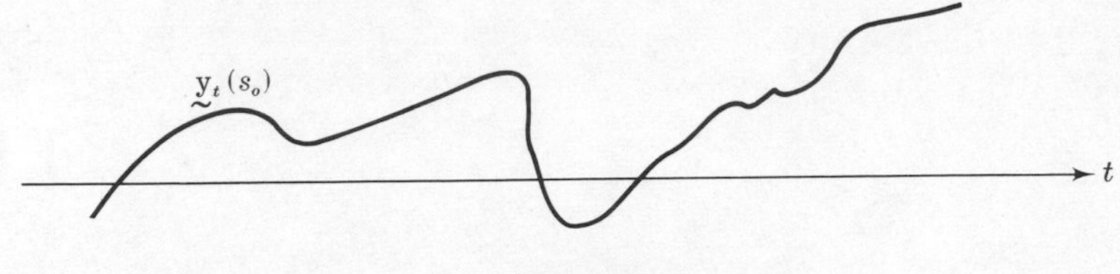

Figure 11.2

As an elementary example of a stochastic process consider a switch which turns on and off at random in time. Assume that the two states are equally likely to begin with. Denote the state of the process at time t by the r.v. $\mathbf{y}_t$; let $\mathbf{y}_t = 1$ or 0 according as the switch is on or off. A realization of the process would appear as shown in Figure 11.3. More precisely assume

(i) $P(\mathbf{y}_0 = 1) = P(\mathbf{y}_0 = 0) = \frac{1}{2}$;

(ii) the probability of n changes of state in the time interval $(t, t + \tau)$ is

$$\frac{\tau^n}{n!} e^{-\tau}; \quad n = 0, 1, 2, \ldots$$

regardless of the state of the process.

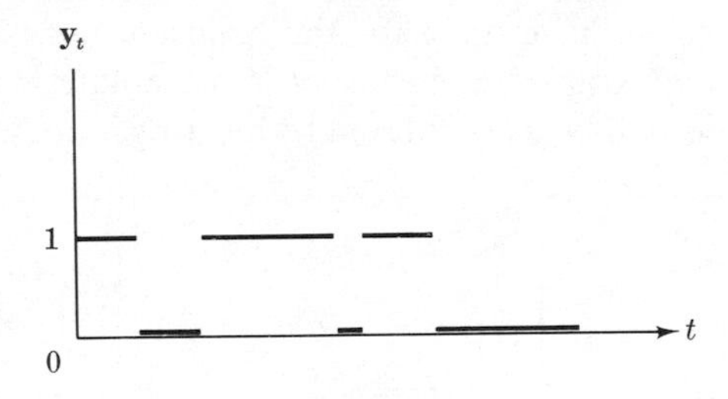

Figure 11.3

These two assumptions determine the probability structure of the process. We have

$$P(\mathbf{y}_t = 1 | \mathbf{y}_0 = 1) = \sum_{n=0}^{\infty} \frac{t^{2n}}{(2n)!} e^{-t} = e^{-t} \cosh t,$$

$$P(\mathbf{y}_t = 1 | \mathbf{y}_0 = 0) = e^{-t} \sinh t,$$

and

$$P(\mathbf{y}_t = 1) = \tfrac{1}{2} e^{-t} \cosh t + \tfrac{1}{2} e^{-t} \sinh t = \tfrac{1}{2}.$$

Thus, "on" and "off" are equally likely and $E\mathbf{y}_t = \frac{1}{2}$ for all t.

In Sections 3.3, 4.7, and 9.6 we have been discussing the *Poisson process* without using that language. For purposes of fixing ideas we have talked about the number of collisions between molecules of a gas which occur over time, but clearly this is a stochastic process in the sense of the present discussion. We have only to imagine that the various molecules in a given volume are indexed by s and let $\mathbf{y}_t(s)$ be the number of collisions up to time t of the molecule labeled s. For fixed t, $\mathbf{y}_t(s)$ is then an r.v. depending on which molecule (determined probabilistically) is chosen; while for fixed $s = s_0$, $\mathbf{y}_t(s_0)$ is a nondecreasing step function showing when the collisions of molecule s_0 take place.

11.2 RENEWAL THEORY[1]

Sections 4.8 and 4.9, on survival functions and car following, can also be considered as introductory to this section. The Poisson process is the prototype for renewal theory but the "no deterioration" property makes it by far the simplest special case.

A *renewal process* is a sequence of r.v.'s $\{\mathbf{y}_1, \mathbf{y}_2, \ldots\}$ of the form

$$\mathbf{y}_n = \mathbf{x}_1 + \cdots + \mathbf{x}_n$$

where $\{\mathbf{x}_1, \mathbf{x}_2, \ldots\}$ are independent and have common d.f. $F(x)$ such that $F(0) = 0$. The time of the nth renewal is $\mathbf{y}_n$; by convention $\mathbf{y}_0 = 0$. $\mathbf{x}_n = \mathbf{y}_n - \mathbf{y}_{n-1}$ is the waiting time between renewals. $\mathbf{N}_t$ is the number of renewals in time t.

For $k \geq 1$, the d.f. of $\mathbf{y}_k$ is $F^{(k)}(x)$, the k-fold convolution of $F(x)$. $\mathbf{N}_t \geq n$ if and only if $\mathbf{y}_n \leq t$. Hence, for $n = 0, 1, \ldots$

$$P(\mathbf{N}_t \geq n) = P(\mathbf{y}_n \leq t) = F^{(n)}(t) \tag{11.1}$$

and

$$P(\mathbf{N}_t = n) = F^{(n)}(t) - F^{(n+1)}(t) \tag{11.2}$$

where, in agreement with our convention, $F^{(0)}(t)$ equals 0 for $t < 0$ and equals 1 otherwise.

Asymptotic results, using the central limit theorem, are easy to obtain for renewal processes. If the mean and variance of waiting time are μ and σ^2 then

$$\lim_{n\to\infty} P\left(\frac{\mathbf{y}_n - n\mu}{\sqrt{n}\sigma} < y\right) = \int_{-\infty}^{y} n(x; 0, 1)\, dx. \tag{11.3}$$

That is, renewal time is asymptotically normal.

[1] For a more detailed treatment see Cox, D. R., *Renewal Theory*. Methuen and Co. Ltd., London, 1962.

Also, writing $n_t = t/\mu + z\sigma\sqrt{t/\mu^3}$,

$$P(\mathbf{N}_t < n_t) = P(\mathbf{y}_{n_t} > t) = P\left(\frac{\mathbf{y}_{n_t} - n_t\mu}{\sqrt{n_t}\sigma} > \frac{t - n_t\mu}{\sqrt{n_t}\sigma}\right).$$

Fixing z but taking $t \to \infty$

$$\lim_{t\to\infty} P(\mathbf{N}_t < n_t) = \lim_{t\to\infty} P\left(\frac{\mathbf{y}_{n_t} - n_t\mu}{\sqrt{n_t}\sigma} > -z\right) = \int_{-\infty}^{z} n(x; 0, 1)\, dx. \quad (11.4)$$

Thus, $\mathbf{N}_t$ is for large t approximately $N(t/\mu, t\sigma^2/\mu^3)$.

Renewal theory is particularly concerned with the *renewal function*

$$\begin{aligned} H(t) = E\mathbf{N}_t &= \sum_{n=0}^{\infty} nP(\mathbf{N}_t = n) \\ &= \sum_{n=1}^{\infty} n\{F^{(n)}(t) - F^{(n+1)}(t)\} \\ &= \sum_{n=1}^{\infty} F^{(n)}(t). \end{aligned} \quad (11.5)$$

Here we concentrate on renewal theory in continuous time. We assume that the waiting times are continuous r.v.'s, the d.f. $F(x)$ admitting a density $f(x)$. The *renewal density* $h(t) = H'(t)$ is given by

$$h(t) = \sum_{n=1}^{\infty} f^{(n)}(t) \quad (11.6)$$

where $f^{(n)}(t)$ is the n-fold convolution of the density $f(x)$. This relation makes it clear that $h(t)\,\Delta t$ is approximately the probability of at least one renewal in the interval $(t, t + \Delta t)$. This is an important interpretation of the renewal density.

We now treat the Poisson process anew from the point of view of renewal theory. The waiting time density is $f(x) = g(x; 1, \rho)$ so that, from Theorem 4.2, $f^{(n)}(x) = g(x; n, \rho)$. Hence,

$$h(t) = \sum_{n=1}^{\infty} \frac{\rho(\rho x)^{n-1}}{\Gamma(n)} e^{-\rho x} = \rho$$

and $H(t) = \rho t$. Also, from Problem 16 of Chapter 4.

$$P(\mathbf{N}_t = n) = \frac{(\rho t)^n}{n!} e^{-\rho t}; \quad n = 0, 1, 2, \ldots. \quad (11.7)$$

The following is one more result concerning the Poisson process; it depends on the above equation and provides an illustration of a conditional joint density function.

Theorem 11.1

Given, for a Poisson process, that N renewals have occurred in time T; then the (unordered) renewal times are independent and uniformly distributed each with density $u(x; 0, T)$.

Proof

The joint density of the waiting times $\mathbf{x}_1, \ldots, \mathbf{x}_N$ given $\mathbf{N}_T = N$ is

$$\frac{\rho e^{-\rho x_1} \cdots \rho e^{-\rho x_N} \cdot e^{-\rho(T - \sum_1^N x_i)}}{P(\mathbf{N}_T = N)} = \frac{\rho^N e^{-\rho T}}{\frac{(\rho T)^N}{N!} e^{-\rho T}} = \frac{N!}{T^N}$$

for $x_i \geq 0$ and $\sum_1^N x_i \leq T$.

Now $\mathbf{y}_1 = \mathbf{x}_1$, $\mathbf{y}_2 = \mathbf{x}_1 + \mathbf{x}_2, \ldots, \mathbf{y}_N = \mathbf{x}_1 + \cdots + \mathbf{x}_N$ and the Jacobian of this transformation is 1. The joint density of the ordered renewal times is therefore

$$f(y_1, \ldots, y_N | N) = \frac{N!}{T^N}$$

over the range $0 \leq y_1 \leq \cdots \leq y_N \leq T$. To find the density $h(t_1, \ldots, t_N | N)$ of the unordered renewal times we observe that, because of symmetry,

$$N!\, h(t_1, \ldots, t_N | N) = f(y_1, \ldots, y_N | N).$$

Hence,

$$h(t_1, \ldots, t_N | N) = T^{-N}, \quad 0 \leq t_i \leq T$$

but this is the product of N uniform densities.

The rest of our discussion of renewal processes employs the Laplace transform. For distributions with a density function $f(x)$, the Laplace transform is

$$f^*(s) = \int_0^\infty e^{-sx} f(x)\, dx.$$

Laplace transforms can be used in much the same way as characteristic functions. The Laplace transform of the density of the sum of two independent r.v.'s is the product of the transforms of the individual densities. Formal expansion of the Laplace transform in a power series is possible and moments may be determined by this method. In particular if μ and σ^2 are mean and variance of the density $f(x)$ then

$$f^*(s) = 1 - s\mu + \tfrac{1}{2}s^2(\sigma^2 + \mu^2) + 0(s^2). \tag{11.8}$$

[$0(s^2)$ is a quantity such that $0(s^2)/s^2$ remains bounded as $s \to 0$.] Also, there is an inversion formula and uniqueness theorem for Laplace transforms of continuous functions.

Two specific formulas will be needed. First, if $f^*(s)$ is the Laplace transform of a density function $f(x)$, then $F^*(s) = f^*(s)/s$ is the Laplace transform of the corresponding distribution function $F(x)$. Second, the Laplace transform of x^p is $\Gamma(p + 1)/s^{p+1}$ for $p > -1$. Formally related to this last result is an asymptotic correspondence between a function and its transform. If, for $s \to 0$,

$$k^*(s) = \frac{A}{s^2} + \frac{B}{s} + 0(1)$$

then, as $x \to \infty$,

$$k(x) = Ax + B + 0(1).$$

We take this last result to be well beyond the scope of our book.

Now from Equation (11.5) the transform of the renewal function is

$$H^*(s) = \frac{1}{s}\sum_{n=1}^{\infty} [f^*(s)]^n = \frac{f^*(s)}{s[1 - f^*(s)]}.$$

And from Equation (11.8), by division of series

$$H^*(s) = \frac{1}{s^2\mu} + \frac{1}{s}\frac{\sigma^2 - \mu^2}{2\mu^2} + 0(1)$$

as $s \to 0$, thus as $t \to \infty$

$$H(t) = \frac{t}{\mu} + \frac{\sigma^2 - \mu^2}{2\mu^2} + 0(1). \tag{11.9}$$

Similarly

$$\lim_{t \to \infty} h(t) = \mu^{-1}. \tag{11.10}$$

Consider now an altered model in which an ordinary renewal process is operating but observation does not begin at a renewal. If the process has been operating for a long time before observation begins then the resulting new process will be called an *equilibrium renewal process*; it will be just like an ordinary renewal process except that the distribution of $\mathbf{x}_1$, the first waiting time, will be altered. We now wish to calculate the equilibrium distribution of this first waiting time.

In an ordinary process let $\mathbf{w}_t$ be the waiting time from the fixed time t to the next renewal. The limiting distribution of $\mathbf{w}_t$, as $t \to \infty$, will be the distribution of $\mathbf{x}_1$ for the equilibrium renewal process. Thus, in an ordinary process, if $\mathbf{w}_t = w$, then either the first renewal occurs at time $t + w$ or for some u a renewal occurs at time $t - u$ and the next renewal occurs $u + w$ time units later. Therefore, the density of $\mathbf{w}_t$ is

$$f(t + w) + \int_0^t h(t - u)f(u + w)\, du.$$

Taking the limit as $t \to \infty$ we see that the density of $\mathbf{x}_1$ in an equilibrium renewal process is

$$f_{\mathbf{x}_1}(w) = \frac{1}{\mu}\int_0^\infty f(u + w)\,du = \frac{1}{\mu}\int_w^\infty f(u)\,du = \frac{\bar{F}(w)}{\mu}. \tag{11.11}$$

Note that this density does in fact integrate to 1.

In summary, an equilibrium renewal process is a sequence of r.v.'s $\{\mathbf{y}_1, \mathbf{y}_2, \ldots\}$ of the form

$$\mathbf{y}_n = \mathbf{x}_1 + \cdots + \mathbf{x}_n$$

where $\{\mathbf{x}_1, \mathbf{x}_2, \ldots\}$ are independent and $\mathbf{x}_2, \mathbf{x}_3, \ldots$ have common d.f. $F(x)$ such that $F(0) = 0$ but $\mathbf{x}_1$ has density $\bar{F}(x)/\mu$.

11.3 THE NORMAL DIFFUSION PROCESS

Here we treat one process related to the multivariate normal distribution. Let $\mathbf{e}_{t,\tau}$ be the increment over the time interval $(t, t + \tau)$ of a family of r.v.'s $\{\mathbf{y}_t\}$, that is, $\mathbf{y}_\tau = \mathbf{y}_t + \mathbf{e}_{t,\tau-t}$. Assume that increments over nonoverlapping intervals are independent. Define

$$F(y; t, \tau) = P(\mathbf{e}_{t,\tau-t} \leq y)$$

and

$$f(y; t, \tau) = \frac{\partial F}{\partial y}.$$

From the equation

$$\mathbf{e}_{t,\tau+\Delta\tau-t} = \mathbf{e}_{t,\tau-t} + \mathbf{e}_{\tau,\Delta\tau}$$

we have by the convolution formula

$$F(y; t, \tau + \Delta\tau) = \int F(y - h; t, \tau) f(h; \tau, \tau + \Delta\tau)\,dh.$$

$$F(y; t, \tau + \Delta\tau) - F(y; t, \tau) = \int [F(y - h; t, \tau) - F(y; t, \tau)] f(h; \tau, \tau + \Delta\tau)\,dh$$

$$\begin{aligned}
\Delta\tau\left[\frac{\partial F}{\partial \tau} + o(\Delta\tau)\right] &= \int_{|h|<\delta} [F(y - h; t, \tau) - F(y; t, \tau)] f(h; \tau, \tau + \Delta\tau)\,dh \\
&= -\frac{\partial F(y; t, \tau)}{\partial y}\int_{|h|<\delta} h f(h; \tau, \tau + \Delta\tau)\,dh \\
&\quad + \frac{1}{2}\frac{\partial^2 F(y; t, \tau)}{\partial y^2}\int_{|h|<\delta} h^2 f(h; \tau, \tau + \Delta\tau)\,dh \\
&\quad + \int_{|h|<\delta} o(h^2) f(h; \tau, \tau + \Delta\tau)\,dh.
\end{aligned}$$

Taking the limit as $\Delta\tau \to 0$,

$$\frac{\partial F}{\partial \tau} = -a(\tau)\frac{\partial F}{\partial y} + \frac{b(\tau)}{2}\frac{\partial^2 F}{\partial y^2} \tag{11.12}$$

where

$$a(\tau) = \lim_{\Delta\tau \to 0} \frac{1}{\Delta\tau}\int h \cdot f(h;\, \tau, \tau + \Delta\tau)\, dh = \lim_{\Delta\tau \to 0} \frac{E\mathbf{e}_{\tau,\Delta\tau}}{\Delta\tau}$$

$$b(\tau) = \lim_{\Delta\tau \to 0} \frac{1}{\Delta\tau}\int h^2 \cdot f(h;\, \tau, \tau + \Delta\tau)\, dh = \lim_{\Delta\tau \to 0} \frac{E(\mathbf{e}_{\tau,\Delta\tau})^2}{\Delta\tau}.$$

Here we have assumed that the limits $a(\tau)$ and $b(\tau)$ exist and that

$$\lim_{\Delta\tau \to 0} \frac{1}{\Delta\tau}\int_{|h| > \delta} f(h;\, \tau, \tau + \Delta\tau)\, dh = 0, \quad \delta > 0.$$

The last assumption is a strong form of stochastic continuity.

$\int_{|h| > \delta} f(h;\, \tau, \tau + \Delta\tau)\, dh$ is the probability of a jump greater than δ in time $\Delta\tau$; not only must this probability approach 0, but it must go to 0 faster than $\Delta\tau$.

The remainder of our work is a study of the important equation (11.12). Consider first the case $a(\tau) = 0$ and $b(\tau) = 1$; we seek d.f.'s $F(y;\, t, \tau)$ which satisfy

(i) $$\frac{\partial F}{\partial \tau} = \frac{1}{2}\frac{\partial^2 F}{\partial y^2}$$

(ii) $$\lim_{y \to \infty} F = 1, \quad \lim_{y \to -\infty} F = 0$$

(iii) $$\lim_{y \to \infty} \frac{\partial F}{\partial y} = \lim_{y \to -\infty} \frac{\partial F}{\partial y} = 0$$

and the stochastic continuity condition

(iv) $$\lim_{\tau \to t} F(y;\, t, \tau) = \begin{cases} 0 & \text{for } y < 0 \\ 1 & \text{for } y \geq 0. \end{cases}$$

We may argue as follows that the solution to the above four relations is unique. Assume that F_1 and F_2 are two solutions. Then $W = F_1 - F_2$ satisfies (i) and (iii) while (ii) and (iv) are replaced by

$$\lim_{y \to \infty} W = \lim_{y \to -\infty} W = 0$$

and

$$\lim_{\tau \to t} W = 0.$$

Define

$$J(t, \tau) = \int_{-\infty}^{\infty} W^2(y; t, \tau)\, dy.$$

Now

$$\frac{\partial J}{\partial \tau} = 2\int_{-\infty}^{\infty} W \frac{\partial W}{\partial \tau}\, dx = \int_{-\infty}^{\infty} W \frac{\partial^2 W}{\partial y^2}\, dy.$$

Integrating by parts,

$$\frac{\partial J}{\partial \tau} = W \frac{\partial W}{\partial y}\Bigg|_{-\infty}^{\infty} - \int_{-\infty}^{\infty} \left(\frac{\partial W}{\partial y}\right)^2 dy$$

the first term on the right is zero and $\partial J/\partial \tau \leq 0$.

$$\frac{J(t, \tau) - J(t, t + 0)}{\tau - t} = \frac{J(t, \xi)}{\partial \tau}; \quad t < \xi \leq \tau.$$

Therefore, since $J(t, t + 0) = 0$,

$$J(t, \tau) = (\tau - t)\frac{\partial J(t, \xi)}{\partial \tau} \leq 0.$$

But $J(t, \tau) \geq 0$ by definition, so that $J(t, \tau) = 0$, $W(y; t, \tau) = 0$, and $F_1 = F_2$.

Next, we may verify directly that

$$f(y; t, \tau) = \frac{1}{\sqrt{2\pi(\tau - t)}} e^{-y^2/2(\tau - t)}$$

satisfies $\partial f/\partial \tau = \frac{1}{2}\partial^2 f/\partial y^2$. Upon integrating this equation

$$F(y; t, \tau) = \int_{-\infty}^{y} \frac{1}{\sqrt{2\pi(\tau - t)}} e^{-x^2/2(\tau - t)}\, dx$$

is found to satisfy (i). We see directly that F also satisfies (ii), (iii), and (iv), and hence, is the unique solution to these four equations.

The general case with arbitrary functions $a(\tau)$ and $b(\tau)$ may be reduced to the special situation $a(\tau) = 0$ and $b(\tau) = 1$. This is done by substituting in (10.12)

$$y^* = y - \int_t^\tau a(z)\, dz$$

$$t^* = \int_0^t b(z)\, dz, \quad \tau^* = \int_0^\tau b(z)\, dz.$$

This gives

$$\frac{\partial F}{\partial \tau} = \frac{\partial F}{\partial \tau^*}\frac{d\tau^*}{d\tau} + \frac{\partial F}{\partial y^*}\frac{\partial y^*}{\partial \tau}$$

$$= \frac{\partial F}{\partial \tau^*} b(\tau) - \frac{\partial F}{\partial y^*} a(\tau)$$

$$\frac{\partial F}{\partial y} = \frac{\partial F}{\partial y^*}\frac{\partial y^*}{\partial y} = \frac{\partial F}{\partial y^*}$$

and

$$\frac{\partial^2 F}{\partial y^2} = \frac{\partial\left(\dfrac{\partial F}{\partial y^*}\right)}{\partial y^*}\frac{\partial y^*}{\partial y} = \frac{\partial^2 F}{\partial y^{*2}}$$

so that (10.12) reduces to (i).

In summary,

$$F(y; t, \tau) = \int_{-\infty}^{y} \frac{1}{\sqrt{2\pi}\,\sigma} e^{-(x-\mu)^2/2\sigma^2}\,dx$$

$$\mu = \int_t^\tau a(z)\,dz \quad \text{and} \quad \sigma^2 = \int_t^\tau b(z)\,dz$$

is the only family of d.f.'s which will describe a process satisfying the imposed conditions. The function $a(\tau)$ may be interpreted as the instantaneous expected rate of change in the process; $b(\tau)$ has a similar interpretation involving squares of increments. The essential restricting conditions were (a) increments over nonoverlapping intervals are independent and (b) the process is continuous in the sense explained earlier.

PROBLEMS

1. Prove that the Laplace transform of x^p is $\Gamma(p+1)/s^{p+1}$ for $p > -1$.
2. Prove that if $f^*(s)$ is the Laplace transform of the density function $f(x)$ then $F^*(s) = f^*(s)/s$ is the Laplace transform of the corresponding distribution function.
3. From (11.4) show that as $t \to \infty$, $\mathbf{N}_t/t \to \mu^{-1}$ in probability.
4. For the "on-off" process of Section 11.1 show that

$$\frac{1}{T}\int_0^T \mathbf{y}_t\,dt \to \frac{1}{2} = E\mathbf{y}_t$$

in probability. A result of this type, where an average over time is shown equal to the expectation at a fixed time, is called an Ergodic theorem.

5. In the notation of Section 11.8, show that

$$\lim_{\Delta\tau \to 0} \int_{|h| > \delta} f(h; \tau, \tau + \Delta\tau)\, dh = 0$$

for $\delta > 0$ is equivalent to

$$\lim_{\tau \to t} F(y; t, \tau) = \begin{cases} 0, & y < 0 \\ 1, & y \geq 0. \end{cases}$$

6. Consider m white balls in an urn to begin with. At each of n trials a ball is chosen at random and removed from the urn; a black ball is then introduced. Let $f_m(y, t) = P$ (proportion of white balls in urn at time t is y) where $t = n/m$. Derive a recursion relationship for $f_m(y, t + 1/m)$ and from this derive a partial differential equation for the limit of f_m as $m \to \infty$. (Assume the limit exists.)

7. Automobiles pass a shop according to a temporal Poisson process with parameter λ; they stop at the shop with constant probability p. Calculate the probability that n cars stop in unit time.

8. A small company has two cars which it rents out by the day. Suppose that demands for a car occur according to a Poisson process with six-hour mean waiting time between demands.
   (a) On what proportion of days is the demand in excess of the capacity (2 cars)?
   (b) On what proportion of days are both cars sitting in the garage?

9. A set of n light bulbs burns continuously. The life of each bulb has a probability density function $\lambda^2 x \exp(-\lambda x)$ so that it may be supposed to consist of the sum of two exponential variables each with mean λ^{-1}. The cost of replacing the whole set is nc_1 and the cost of replacing one bulb as it fails is $c_2 (c_1 < c_2)$. A policy of periodical total replacement is followed and individual bulbs are replaced as they fail. Select the best interval between complete replacements.

10. Let $\mathbf{e}_{t,\tau}$ be the increment over the interval $(t, t + \tau)$ of a family of r.v.'s $\{\mathbf{x}_t\}$, that is, $\mathbf{x}_{t+\tau} = \mathbf{x}_t + \mathbf{e}_{t,\tau}$. Assume (a) that increments over nonoverlapping intervals are independent, (b) the distribution of $\mathbf{e}_{t,\tau}$ depends only on τ, (c) the expectation and variance of $\mathbf{e}_{t,\tau}$ exist and are continuous functions of τ. Show that if $\mathbf{x}_0 = 0$ then the mean and variance of $\mathbf{e}_{t,\tau}$ are $m\tau$ and $a\tau$ where $m \geq 0$ and $a \geq 0$.

11. (continuation) In the previous problem take $m = 0$ and let $\phi_\tau(s)$ be the c.f. of $\mathbf{e}_{t,\tau}$. Now

$$\mathbf{e}_{t,\tau} = \mathbf{e}_{t,\tau/n} + \mathbf{e}_{t+\tau/n,\tau/n} + \cdots + \mathbf{e}_{t-\tau/n,\tau/n}.$$

Note that the variables in this sum are independent, identically distributed, have mean zero, and have variance $a\tau/n$. As in the proof of the central limit theorem

$$\phi_\tau(s) = [\phi_{\tau/n}(s)]^n \simeq \left(1 - \frac{a\tau}{n}\frac{s^2}{2}\right)^n \to e^{-a\tau s^2/2}$$

as $n \to \infty$. Thus $\mathbf{e}_{t,\tau}$ is $N(0, a\tau)$. Locate the error in this argument and provide a counterexample.

NOTES ON PROBLEMS

CHAPTER 4

12. Consider the degenerate distribution obtained by placing a unit of mass on the dotted line shown in Figure A.

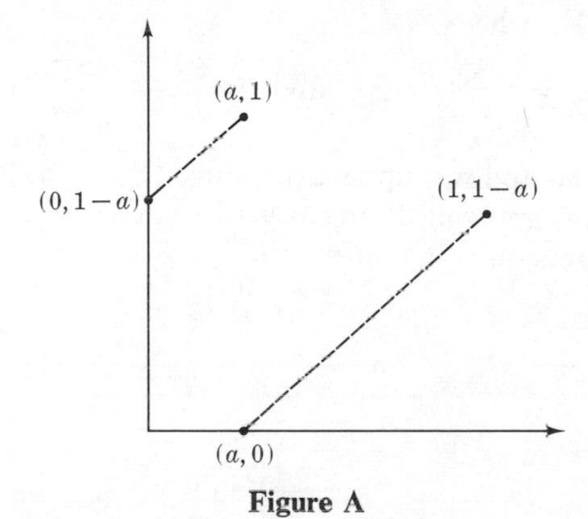

Figure A

CHAPTER 9

3. Proof

The second hypothesis can be written as $\psi_{m,n}(t, s) \to \psi(t, s)$, uniformly in every sphere, where $\psi_{m,n}$ and ψ are appropriately defined characteristic functions. Now since $\psi(t, s)$ is uniformly continuous

$$\psi_{m,n}\left[t\left(1 + \frac{c_n{}^2}{a_m{}^2}\right)^{-1/2}, -t\left(1 + \frac{a_m{}^2}{c_n{}^2}\right)^{-1/2}\right] \to \psi(t(1 + \lambda^2)^{-1/2}, -t(1 + \lambda^{-2})^{-1/2})$$

which is the characteristic function of $\mathbf{f}^*(1 + \lambda^2)^{-1/2} - \mathbf{g}^*(1 + \lambda^{-2})^{-1/2}$.

8. Proof

The characteristic function

$$\int_{-\infty}^{\infty}\cdots\int R(y) \exp\left[-\tfrac{1}{2}(1 - 2it)H(y)\right] dy_1 \cdots dy_p$$

$$= (1 - 2it)^{-(m+p)/q} \int_{-\infty}^{\infty}\cdots\int R(w) \exp\left[-\tfrac{1}{2}H(w)\right] dw_1 \cdots dw_p$$

$$= (1 - 2it)^{-(m+p)/q},$$

where we have made the transformation

$$w_j = (1 - 2it)^{1/q} y_j.$$

But $(1 - 2it)^{-(m+p)/q}$ is the characteristic function of the chi-square distribution with $2(m + p)/q$ degrees of freedom, and hence, $H(x)$ must have this distribution.

10. To prove this result where the c's may have arbitrary sign we write $\sum c_i y_i^2 = \Sigma_+ - \Sigma_-$ where

$$\Sigma_+ = \sum_{\{i:c_i \geq 0\}} c_i y_i^2 \quad \text{and} \quad -\Sigma_- = \sum_{\{i:c_i \geq 0\}} c_i y_i^2.$$

The equation then follows upon expanding $(\Sigma_+ - \Sigma_-)^j$ according to the binomial theorem, applying Problem 9 to each term and then regrouping into a single expression.

CHAPTER 11

6.

$$f_m\left(y, t + \frac{1}{m}\right)$$

$$= P(\text{black}|\text{prop. is } y) f_m(y, t) + P\left(\text{white}|\text{prop. is } y + \frac{1}{m}\right) f_m\left(y + \frac{1}{m}, t\right)$$

$$f_m\left(y, t + \frac{1}{m}\right) = (1 - y) f_m(y, t) + \left(y + \frac{1}{m}\right) f_m\left(y + \frac{1}{m}, t\right)$$

$$\frac{f_m\left(y, t + \frac{1}{m}\right) - f_m(y, t)}{1/m} = y \frac{f_m\left(y + \frac{1}{m}, t\right) - f_m(y, t)}{1/m} + f_m\left(y + \frac{1}{m}, t\right)$$

$$\frac{\partial f(y, t)}{\partial t} = y \frac{\partial f(y, t)}{\partial y} + f(y, t).$$

Table of the Standard or $N(0, 1)$ Distribution Function

$$N(x) = \int_{-\infty}^{x} (2\pi)^{-1/2} \exp\left(-\frac{v^2}{2}\right) dv$$

x	$N(x)$	x	$N(x)$
0.00	0.50000	2.05	0.97982
0.05	0.51994	2.10	0.98214
0.10	0.53983	2.15	0.98422
0.15	0.55962	2.20	0.98610
0.20	0.57926	2.25	0.98778
0.25	0.59871	2.30	0.98928
0.30	0.61791	2.35	0.99061
0.35	0.63683	2.40	0.99180
0.40	0.65542	2.45	0.99286
0.45	0.67364	2.50	0.99379
0.50	0.69146	2.55	0.99461
0.55	0.70884	2.60	0.99534
0.60	0.72575	2.65	0.99598
0.65	0.74215	2.70	0.99653
0.70	0.75804	2.75	0.99072
0.75	0.77337	2.80	0.99744
0.80	0.78814	2.85	0.99781
0.85	0.80234	2.90	0.99813
0.90	0.81594	2.95	0.99841
0.95	0.82894	3.00	0.99865
1.00	0.84134	3.05	0.99886
1.05	0.85314	3.10	0.99903
1.10	0.86433	3.15	0.99918
1.15	0.87493	3.20	0.99931
1.20	0.88493	3.25	0.99942
1.25	0.89435	3.30	0.99952
1.30	0.90320	3.35	0.99960
1.35	0.91149	3.40	0.99966
1.40	0.91924	3.45	0.99972
1.45	0.92647	3.50	0.99977
1.50	0.93319	3.55	0.99981
1.55	0.93943	3.60	0.99984
1.60	0.94520	3.65	0.99987
1.65	0.95053	3.70	0.99989
1.70	0.95543	3.75	0.99991
1.75	0.95994	3.80	0.99993
1.80	0.96407	3.85	0.99994
1.85	0.96784	3.90	0.99995
1.90	0.97128	3.95	0.99996
1.95	0.97441	4.00	0.99997
2.00	0.97725	4.05	0.99997

INDEX

AMBUSH!

A Robin Hood Adventure

Essex County Council
30130212428147

ReadZone Books Limited

www.ReadZoneBooks.com

© in this edition 2016 ReadZone Books Limited

This print edition published in cooperation with Fiction Express, who first published this title in weekly instalments as an interactive e-book.

FICTION EXPRESS

Fiction Express
First Floor Office, 2 College Street,
Ludlow, Shropshire SY8 1AN
www.fictionexpress.co.uk

Find out more about Fiction Express on pages 57–58.

Design: Laura Harrison & Keith Williams
Cover Image: Bigstock

© in the text 2016 Sara Vogler and Jan Burchett
The moral right of the author has been asserted.

All rights reserved. No part of this publication may be reproduced, stored in a retrieval system or transmitted, in any form or by any means, electronic, mechanical, photocopying, recording or otherwise, without the prior permission of ReadZone Books Limited.

ISBN 978-1-78322-599-6

Printed in Malta by Melita Press